T0332917

Edited by

Progress in Physics
Vol. 8

Edited by
A. Jaffe, G. Parisi,
and D. Ruelle

Birkhäuser
Boston · Basel · Stuttgart

Workshop on Non-Perturbative Quantum Chromodynamics

Kimball A. Milton,
Mark A. Samuel, editors

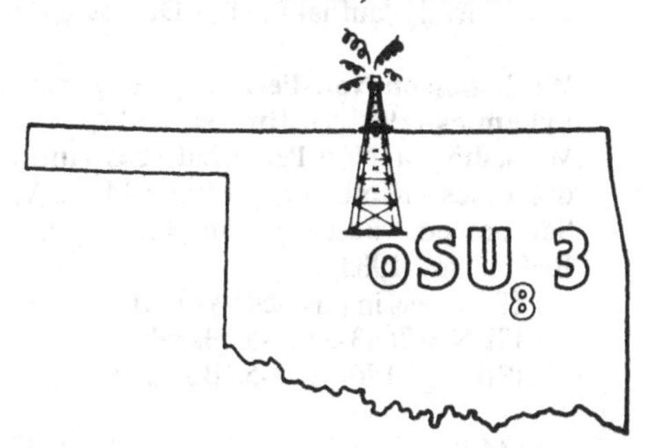

1983

Birkhäuser
Boston • Basel • Stuttgart

Editors:

Kimball A. Milton
Mark A. Samuel
Department of Physics
Oklahoma State University
Stillwater, OK 74078

Library of Congress Cataloging in Publication Data

Workshop on Non-perturbative Quantum Chromodynamics
(1983 : Oklahoma State University)
Workshop on Non-perturbative Quantum Chromodynamics.

(Progress in physics ; vol. 8)
1. Quantum chromodynamics—Congresses. I. Milton,
K. A. II. Samuel, Mark A. III. Title. IV. Series:
Progress in physics (Birkhäuser Boston) ; vol. 8.
QC793.3.Q35W67 1983 539.7'21 83-22521
ISBN 0-8176-3127-5

CIP-Kurztitelaufnahme der Deutschen Bibliothek

Workshop on Non-Perturbative Quantum Chromodynamics (1983, Stillwater, Okla.):
Workshop on Non-Perturbative Quantum Chromodynamics / Kimball A. Milton ; Mark A.
Samuel, ed. - Basel ; Stuttgart ; Boston :
Birkhäuser, 1983.
(Progress in physics ; Vol. 8)
ISBN 3-7643-3127-5 (Basel ...)
ISBN 0-8176-3127-5 (Boston)

NE: Milton, Kimball A. [Hrsg.]; HST; GT

ISBN 0-8176-3127-5
ISBN 3-7643-3127-5
Printed in USA

9 8 7 6 5 4 3 2 1

"A scientist is a man that can find out anything, and nobody in the world has any way of proving whether he ever found it out or not, and the more things he can think of that nobody can find out about...why, the bigger the scientist he is."

Will Rogers

1. Doug McKay, Kansas
2. Louise Dolan, Rockefeller
3. Dennis Sivers, Argonne
4. Wally Greenberg, Maryland
5. Thomas Morgan, Nebraska
6. Pat Skubic, Oklahoma
7. Gordon Lasher, IBM
8. Bindu Bambah, Chicago
9. Ronald Kantowski, Oklahoma
10. Kim Milton, OSU
11. Gary Köhler, Oklahoma
12. Mark Samuel, OSU
13. Keith Andrew, Arkansas
14. Richard Woloshyn, TRIUMF
15. Mike Lieber, Arkansas
16. Carl Bender, Washington U.
17. P.K. Williams, DOE
18. Charles Nelson, SUNY at Binghamton
19. Steve Adler, Institute for Advanced Study
20. Achin Sen, OSU
21. Jung Rno, Cincinnati
22. Janos Polonyi, Illinois
23. Yoichi Kazama, Kyoto
24. Pahol Sanganetra, Kansas
25. Thomas Kruecken, Nebraska
26. Q. Hokim, Laval University
27. Porter Johnson, IIT
28. Mike Cornwall, UCLA
29. Ken Johnson, MIT
30. Jim Ball, Utah
31. Amarjit Soni, UCLA
32. Steve Pinsky, Ohio State
33. James Reid, Simon Fraser
34. Bruce Hudson, Oklahoma
35. Stan Brodsky, SLAC
36. Seljuk Saritepe, OSU
37. F.C. Chang, St. John's
38. Maria Teresa Thomaz, Wisconsin
39. Michael Creutz, Brookhaven
40. Carl Shakin, Brooklyn College
41. Sue Willis, Oklahoma
42. Adolpho Nimirovsky, Kansas
43. Ted Grose, OSU
44. Unidentified
45. Richard Haymaker, LSU
46. Herman Munczek, Kansas
47. Thomas Koppel, Chalk River
48. Gerrit Schierholz, Hamburg
49. Gary Tupper, OSU
50. Walter Wilcox, OSU
51. Morten Laursen, OSU
52. H.C. Lee, Chalk River
53. Garrett Sylvester, OSU

(OSU = Oklahoma State University)

"FORWARD"

The workshop on Non-Perturbative Quantum Chromodynamics took place over a three-day period, March 7-9, 1983 on the campus of Oklahoma State University in Stillwater. Some 65 participants and 28 talks represented diverse institutions and a broad spectrum of approaches to the central problems of strong interaction physics. Many subjects were discussed, including lattice gauge theory, infrared behavior, confinement, chromostatics and glueballs, as well as the interface with perturbative QCD. The excellent talks and the lively discussions which took place at the sessions, as well as during the various social events, ensured the success of the workshop.

We would like to summarize some of the highlights of those 3 days not represented by the technical papers included here. The workshop opened with a champagne reception at the Student Union Hotel and a party was held at the Samuel's residence Monday evening. A reception and banquet took place Tuesday evening followed by 3 significant events:
(1) A unique feature of this workshop was the contest held for the best interpretation of the workshop logo. The entries were extremely creative and imaginative. After some difficulty, a winner was chosen and Steve Adler became the very proud owner of a brand new OSU orange cowboy hat. (See appendix for some of the entries.)
(2) P.K. Williams of the Department of Energy gave us an up-to-the-minute review of the Federal funding situation especially in regard to CBA. His after dinner talk "Potomac Fever" was both amusing and informative.
(3) The participants enjoyed a taste of the old west in the entertainment provided by the Prairie Dance Theatre Company of Oklahoma City. They presented a private performance of "All Our Yesterdays - Belle Starr", a modern dance look at a portion of early Oklahoma history not too far removed from us in time.

We are indebted to many people who helped assure the smooth functioning and success of the workshop: Smith Holt, Dean of the College of Arts and Sciences who gave the opening welcome; Stan Green and Stan Dunham of Arts and Sciences Extension; Jeff Carroll, Said Ghnein, Ted Grose, Morten Laursen, Michele Plunkett, Seljuk Saritepe, Achin Sen, Geoff Summers, N.V.V.J. Swamy, Gary Tupper, Walter Wilcox, Tim Wilson and Laura Woodward of the Department of Physics; Beth Shumway of the Prairie Dance Theatre Company; and the efficient staffs of Arts and Sciences Extension, the Student Union and the Student Union Hotel. Special thanks go to our wives Margarita Baños-Milton and Carol Samuel for help and moral support.

We are also indebted to the U.S. Department of Energy, the National Science Foundation and the College of Arts and Sciences of Oklahoma State University for financial support.

Kim Milton
Mark Samuel
Editors

TABLE OF CONTENTS

** manuscript not available

PROBLEMS AT THE INTERFACE BETWEEN PERTURBATIVE AND NONPERTURBATIVE QUANTUM CHROMODYNAMICS*

STANLEY J. BRODSKY

Stanford Linear Accelerator Center

Stanford University, Stanford, California 94305

Written in Collaboration with

G. T. BODWIN

Argonne National Laboratory

Argonne, Illinois 60439

and

G. PETER LEPAGE

Laboratory of Nuclear Studies

Cornell University, Ithaca, New York 14853

1. Introduction

Predictions based on perturbative QCD[1] rest on three premises: (1) that hadronic interactions become weak in strength at small invariant separation $r \ll \Lambda_{QCD}^{-1}$; (2) that the perturbative expansion in $\alpha_s(Q)$ is well-defined; and (3) factorization: all effects of collinear singularities, confinement, nonperturbative interactions, and bound state dynamics can be isolated at large momentum transfer in terms of (process independent) structure functions $G_{i/H}(x,Q)$, fragmentation functions $D_{H/i}(z,Q)$, or in the case of exclusive processes, distribution amplitudes $\phi_H(x_i,Q)$.[2]

The assumption of weak hadronic interactions at small separation is consistent with the presumed behavior of confining potentials at short distances, e.g., $V_{conf}(r) \sim \kappa r \leq 50$ MeV for $r < 10^{-15} cm$, and the asymptotic freedom property of perturbative QCD ($\beta_0 = 11 - 2/3\, n_f$):

$$\alpha_s(Q^2) = \frac{4\pi}{\beta_0 \ell n \frac{Q^2}{\Lambda^2}} < 0.2 \quad \text{for } Q > 20\, GeV,\ r < 10^{-15}\, cm\ . \tag{1.1}$$

The assumption that the perturbative expansion for hard scattering amplitudes converges has certainly not been demonstrated; in addition, there are serious ambiguities concerning the choice of renormalization scheme and scale choice Q^2 for

---

* Work supported by the Department of Energy, contract DE-AC03-76SF00515.

the expansion in $\alpha_s(Q^2)$. We will discuss a new procedure[3] to at least partly rectify the latter problem in Section 2.

In the case of exclusive processes, the factorization of hadronic amplitudes at large momentum transfer in the form of distribution amplitudes convoluted with hard scattering quark-gluon subprocess amplitudes can be demonstrated systematically to all orders in $\alpha_s(Q^2)$.[2,4,5] In the case of inclusive reactions, factorization remains an ansatz; general all-orders proofs do not exist[6] because of the complications of soft initial state interactions for hadron-induced processes; thus far factorization has only been verified[7,6] to two loops beyond lowest order in a regime where the applicability of perturbation theory is in doubt. However, we shall show that a necessary condition[6] for the validity of factorization in inclusive reactions is that the momentum transfer must be large compared to the (rest frame) length of the target. We review the present status of the factorization ansatz in Section 3.

The basic form of the factorization ansatz for inclusive reactions at large momentum transfer is[8]

$$d\sigma(\{H\} \to \{H'\} + X) = \sum_{subprocesses} \int \prod_{H,H'} G_{i/H}(x_i, Q)\, D_{H'/j}(z_j, Q) \times d\sigma_{(a)}\left(x_i P_H, \frac{1}{z_j} P_{H'}; Q\right) \prod_{i,j} dx_i dz_j / z_j \tag{1.2}$$

where the structure and fragmentation probability distributions G and D for each hadron is a function of the parton light cone momentum fractions $x = (k^0 + k^z)/(p_H^0 + p_H^z)$ and the summation is an incoherent sum over all leading hard-scattering QCD subprocesses computed with the partons i, j collinear with the initial and final hadron directions. By definition the subprocess cross section has no collinear singularities so it can be expanded in powers of $\alpha_s(Q^2)$. In general the summation includes higher twist power-law suppressed subprocesses where the scattering partons are multiquark, multigluon, hadron or other QCD composite systems.[9,10] The power-law scaling of such processes, which can be important at the edge of phase-space, is determined by dimensional counting.[9,11] In addition one must allow for multiple collision processes,[12] which can be especially important for high transverse energy triggers.

Radiation collinear up to the scale Q is included in the definition of the structure functions, leading to evolution equations and moments of the form:[1]

$$\mathcal{M}^n_{i/H}(Q) = \int_0^1 dx\, x^{n-1} G_{i/H}(x, Q) = \mathcal{M}^n_{i/H}(Q_0) e^{-\gamma_n \xi(Q,Q_0)} \tag{1.3}$$

where

$$\xi(Q,Q_0) = \frac{1}{4\pi}\int_{Q_0^2}^{Q^2} \frac{d\ell^2}{\ell^2}\alpha_s(\ell^2) \simeq \frac{1}{\beta_0}\ell n\left(\frac{\ell n\, Q^2}{\Lambda^2} \Big/ \frac{\ell n\, Q_0^2}{\Lambda^2}\right). \tag{1.4}$$

and γ_n, the standard anomalous dimensions, are independent of the bound state hadron dynamics.

Similarly in the case of exclusive reactions one can isolate the long-distance confinement dynamics from the short-distance quark and gluon hard scattering dynamics — at least to leading order in $1/Q^2$. Hadronic amplitudes take the factorized form[2,4]

$$T = \int \prod_H \phi_H(x_i,Q)\ \hat{T}(x_i,p_H;Q)[dx_i]\ .$$

where $\phi_H(x_i,Q)$ is a universal distribution amplitude which gives the probability amplitude for finding the valence $q\bar{q}$ or qqq in the hadronic wave function collinear up to the scale Q, and $\hat{T}$ is the hard-scattering amplitude for scattering valence quark collisions with the incident and outgoing hadrons.

For example, the pion form factor at large Q^2 takes the form

$$F_\pi(Q^2) = \int_0^1 dx \int_0^1 dy\, \phi^*(x,Q)\ T_H(x,y;Q)\,\phi(y,Q) \tag{1.6}$$

where

$$\phi_\pi(x,Q) = \int^Q \frac{d^2k_\perp}{16\pi^2}\ \psi^Q_{q\bar{q}/\pi}(x,\vec{k}_\perp) \tag{1.7}$$

is the amplitude for finding the q and $\bar{q}$ in the valence state of the pion collinear up to scale Q with light cone longitudinal momentum fractions x and $1-x$, and

$$T_H = \frac{16\pi C_F\alpha_s[Q^2(1-x)(1-y)]}{(1-x)(1-y)Q^2}\left[1+\frac{O[\alpha_s(Q^2)]}{\pi}\right] \tag{1.8}$$

is the probability amplitude for scattering collinear constituents from the initial to the final direction. By definition, T_H contains only transverse momenta greater than Q in loop integrations so that T_H can be expanded in powers of $\alpha_s(Q^2)$. (The superscript Q in $\psi^Q_{q\bar{q}}$ indicates that all internal loop in $\psi_{q\bar{q}}$ are to be cutoff at $k_\perp^2 < Q^2$.) The log Q^2 dependence of the distribution amplitude $\phi(x,Q)$ is determined by the operator product expansion on the light-cone or an evolution equation; its specification at subasymptotic momentum requires the solution to the pion bound state problem. In general we have

$$\phi(x,Q) = x(1-x)\sum_n \phi_n(x,Q_0)\, e^{-\gamma_n \xi(Q,Q_0)} \tag{1.9}$$

where to one loop order, $\phi_n(x,Q_0) = a_n(Q_0)C_n^{3/2}(2x-1)$ are the eigensolutions of the evolution equation, the γ_n are anomalous dimensions[13] analogous to those that appear in Eq. (1.3), and the $a_n(Q_0)$ are determined by the bound state dynamics. The general form of $F_\pi(Q^2)$ is then

$$F_\pi(q^2) = \left|\sum_{n=0}^{\infty} a_n \log^{-\gamma_n}\frac{Q^2}{\Lambda^2}\right|^2 C_F\frac{\alpha_s(Q^2)}{Q^2}\left[1+O\left(\frac{\alpha_s(Q^2)}{\pi}\right)+O\left(\frac{m}{Q}\right)\right]. \tag{1.10}$$

Similar calculations[14] determine the baryon form factors, decay amplitudes such as $\Upsilon \to B\bar{B}$ and fixed angle scattering processes such as Compton scattering, photoproduction, and hadron-hadron scattering, although the latter calculations are complicated by the presence (and suppression) of pinch singularities.[15] It is interesting to note that $\phi(x,Q^2)$ can be measured directly from the angular $\theta_{c.m.}$ dependence of the $\gamma\gamma \to \pi^+\pi^-$ and $\gamma\gamma \to \pi^0\pi^0$ cross sections at large s.[16] In addition, independent of the form of the meson wave function, we can obtain α_s from the ratio[2]

$$\alpha_s(Q^2) = \frac{F_\pi(Q^2)}{4\pi Q^2|F_{\pi\gamma}(Q^2)|^2}\left[1+0\left(\frac{\alpha_s(Q^2)}{\pi}\right)\right] \tag{1.11}$$

where the transition form factor $F_{\pi\gamma}(Q^2)$ can be measured in the two photon reaction $\gamma^*\gamma \to \pi^0$ via $ee \to \pi^0 ee$. Equation (1.11) is in principle one of the cleanest ways to measure α_s [see also Eq. (2.15)]. Higher order corrections in α_s are discussed in Refs. 5,17.

Thus an essential part of the QCD predictions is the hadronic wave functions which determine the probability amplitudes and distributions of the quark and gluons which enter the short distance subprocesses. Computation of the quark and gluon fragmentation function into hadrons require knowledge of the coherent amplitudes which form partons into hadrons. Thus hadronic wave functions provide the link between long distance nonperturbative and short-distance perturbative physics.[18] Eventually, one can hope to compute the wave functions from theory, e.g., from lattice or bag models, or directly from the QCD equations of motion, as we shall outline below. Knowledge of the hadronic wave function also allows the normalization and specification of several types of power law suppressed (higher twist) contributions, such as $1/Q^2$ contributions to the longitudinal structure function of mesons and baryons[17] at $x \to 1$, and direct meson and baryon production subprocesses.[20,21]

The wave function $\psi^{\pi}_{q\bar{q}}(x, k_\perp)$ which appears in Eq. (1.7). is related to the Bethe-Salpeter amplitude at equal "time" $\tau = t + z$ on the light-cone in $A^+ = 0$ gauge.[22] The quark has transverse momentum $k_\perp$ relative to the pion direction and fractional "light-cone" momentum $x = (k^0 + k^3)/(p^0 + p^3) = k^+/p^+$. The state is off the light cone $k^- = k^0 - k^3$ energy shell. In general a hadron state can be expanded in terms of a complete set of Fock state at equal τ:

$$|\pi\rangle = |q\,\bar{q}\rangle\psi_{q\bar{q}} + |q\,\bar{q}\,g\rangle\psi_{q\bar{q}g} + \cdots \tag{1.12}$$

with

$$\sum_n \int [d^2k_\perp]\,[dx]\,|\psi_n(x_i, \vec{k}_{\perp i})|^2 = 1\,.$$

(We suppress helicity labels.) At large Q^2 only the valence state contributes to an exclusive process, since by dimensional counting an amplitude (in a physical gauge) is suppressed by a power of $1/Q^2$ for each constituent required to absorb large momentum transfer. The amplitudes ψ_n are infrared finite for color-singlet bound states. The meson decay amplitude (e.g. $\pi^+ \to \mu^+\nu$) implies a sum rule

$$\frac{a_0}{6} = \frac{1}{2\sqrt{n_c}} f_\pi = \int_0^1 dx\,\phi_\pi(x, Q). \tag{1.13}$$

This result, combined with the constraint on the wave function from $\pi^o \to \gamma\gamma$ requires that the probability that the pion is in its valence state is $\leq 1/4$.[18] Given the $\{\psi_n\}$ for a hadron, virtually any hadronic property can be computed, including anomalous moments, form factors (at any Q^2), etc.

The $\{\psi_n\}$ also determine the basic form of the structure functions appearing in deep inelastic scattering ($a = q, \bar{q}\,, g$)

$$G_{a/p}(x, Q) = \sum_n \int^Q [d^2k_\perp]\,[dx]\,|\psi^Q_n(x_i, k_{\perp i})|^2\,\delta(x - x_i) \tag{1.14}$$

where one must sum over all Fock states containing the constituent a and integrate over all transverse momentum $d^2k_\perp$ and the light-cone momentum fractions $x_i \neq x_a$ of the spectators. The valence state dominates $G_{q/p}(x, Q)$ at the edge of phase space, $x \to 1$. All of the multiparticle x and $k_\perp$ momentum distributions needed for multiquark scattering processes can be defined in a similar manner. The evolution equation for the $G_a(x, Q^2)$ can be easily obtained from the high $k_\perp$ dependence of the perturbative contributions to ψ.

There are many advantages[22] obtained by quantizing a renormalizable local $\tau = t + z$. These field theory at fixed light-cone time include the existence of an orthornormal relativistic wave function expansion, a convenient τ-ordered perturbative theory, and diagonal (number-conserving) charge and current operators. The central reason why one can construct a sensible relativistic wave function Fock state expansion on the light cone is the fact that the perturbative vacuum is also an eigenstate of the full Hamiltonian. The equation of state for the $\{\psi_n(x_i, k_{\perp i})\}$ takes the form

$$H_{LC}\Psi = M^2\Psi \tag{1.15}$$

where

$$H_{LC} = \sum_{i=1}^{n} \left(\frac{k_\perp^2 + m^2}{x}\right)_i + V_{LC} \,, \tag{1.16}$$

and V_{LC} is derived from the QCD Hamiltonian in $A^+ = 0$ gauge quantized at equal τ, and Ψ is a column matrix of the Fock state wave functions. Ultraviolet regularization and invariance under renormalization is discussed in Refs. 2,18.

A comparison of the properties of exclusive and inclusive cross sections in QCD is given in Table I. Given the $\{\psi_n\}$ we can also calculate decay amplitudes, e.g. $\psi \to p\bar{p}$ which can be used to normalize the proton distribution amplitudes. The constraints on hadronic wave functions which result from present experiments are given in Ref. 18. An approximate connection between the valence wave functions defined at equal τ with the rest frame wave function is also given in Refs. 7,8,23, so that one can make predictions from nonperturbative analyses such as bag models, lattice gauge theory, chromostatic approximations, potential models, etc. Other constraints from QCD sum rules are discussed in Ref. 24.

It is interesting to note that the higher twist amplitudes such as $\gamma q \to Mq$, $gq \to Mq$, $q\bar{q} \to M\bar{M}$, $q\bar{q} \to Bq$ which can be numerically important for inclusive hadron production reactions at high $x_\perp$ are absolutely normalized in terms of the distribution amplitudes $\phi_M(x, Q)$, $\phi_B(x_i, Q)$, using the same analysis as that used for form factors. In fact "direct" amplitudes[20,21] such as $\gamma q \to Mq$, $q\bar{q} \to Mg$ and $gq \to Mq$ where the meson interacts directly in the subprocess are rigorously related

Table I. Comparison of exclusive and inclusive cross section

Exclusive Amplitudes	Inclusive Cross Sections
$\mathcal{M} \sim \Pi\, \phi(x_i, Q) \otimes T_H(x_i, Q)$	$d\sigma \sim \Pi\, G(x_a, Q) \otimes d\hat{\sigma}\,(x_a, Q)$
$\phi(x, Q) = \int^Q [d^2k_\perp]\, \psi^Q_{val}(x, k_\perp)$	$G(x, Q) = \Sigma_n \int^Q [d^2k_\perp]\, [dx]'\, \lvert\psi^Q_n(x, k_\perp)\rvert^2$
Measure ϕ in $\gamma\gamma \to M\bar{M}$	Measure G in $\ell p \to \ell X$
$\Sigma_{i\epsilon H}\, \lambda_i = \lambda_H$	$\Sigma_{i\epsilon H}\, \lambda_i \neq \lambda_H$
Evolution	
$\frac{\partial\phi(x,Q)}{\partial \log Q^2} = \alpha_s \int [dy]\, V(x,y)\, \phi(y)$	$\frac{\partial G(x,Q)}{\partial \log Q^2} = \alpha_s \int dy\, P(x/y)\, G(y)$
$\lim_{Q\to\infty} \phi(x, Q) = \Pi_i\, x_i \cdot C_{\text{flavor}}$	$\lim_{Q\to\infty} G(x, Q) = \delta(x)\, C$
Power Law Behavior	
$\frac{d\sigma}{dx}\,(A + B \to C_D) \simeq \frac{1}{s^{n-2}}\, f(\theta_{c.m.})$	$\frac{d\sigma}{d^2p/E}\,(AB \to CX) \simeq \Sigma \frac{(1-x_T)^{2n_s-1}}{(Q^2)^{n_{act}-2}}\, f(\theta_{c.m.})$
$n = n_A + n_B + n_C + n_D$	$n_{act} = n_a + n_b + n_c + n_d$
T_H: expansion in $\alpha_s(Q^2)$	$d\hat{\sigma}$: expansion in $\alpha_s(Q^2)$
Complications	
End point singularities	Multiple scales
Pinch singularities	Phase-space limits on evolution
High Fock states	Heavy quark thresholds
	Heavy twist multiparticle processes
	Initial and final state intractions

to the meson form factor since the same moment of the distribution amplitude appears in each case.

At present there appear to be overwhelming evidence that perturbative QCD provides a viable theory of strong interactions at short distances. The evidence extends from e^+e^- annihilation (the scaling and normalization of $R_{e^+e^-}$, 3-jet events, $\psi \to 3$ jets), $\gamma\gamma$ annihilation ($\gamma\gamma \to$Jets, $F_{2\gamma}(x, Q^2)$), deep inelastic lepton scattering (structure function scaling and evolution), lepton pair production (normalization and scaling behavior, $Q_\perp$ growth), exclusive processes (dimensional counting, relative normalization), large transverse momentum hadron reactions (jets, charge correlations reflecting elementary QCD subprocesses), etc. The most interesting anomalies not readily understood in terms of the standard picture are

1. Charm production in hadronic collisions[25]. The $pp \to$ charm cross sections at ISR energies are much larger ($\sigma_c \sim 1\,mb$) and much flatter in x_L than predicted by the usual gluon fusion model ($gg \to c\bar{c}$). Indications for a significant charm quark distribution ($P_{c\bar{c}} \sim 1\%$) at large $x_{Bj} > 0.4$ increasing with W^2 are also suggested by EMC deep inelastic meson scattering measurements. The possibility that hadronic production of charm can be understood in terms of intrinsic charm states in the hadronic wave function is discussed in Ref. 26.
2. The $pp \to pX$ cross section at FNAL energies[27] scales roughly as $E\, d\sigma/d^3p \sim p_T^{-12} F(x_T, \theta_{cm})$, which is incompatible with the scaling laws predicted by quark fragmentation into protons derived from leading twist subprocesses. The approximate empirical scaling behavior[28] of the γ/π ratio $\sim p_T^2 F(x_T, \theta_{cm})$ also hints at significant higher twist contributions for meson production.
3. EMC and SLAC measurements show that simple additivity $F_{2A}(x, Q^2) = A F_{2N}(x, Q^2)$ for nuclear structure functions breaks down at a significant level ($\pm 20\%$), a much larger deviation than that expected from shadowing and binding effects.[29] A range of possibilities have been suggested to explain this phenomena, such as anti-shadowing mechanisms,[30] anomalous isobar/meson degrees of freedom in the nucleus,[31] or physical changes of the nucleon quark wave function due to the nuclear environment.[32]

The above experimental anomalies do not really conflict with the basic premise that QCD is the correct theory of hadron interactions. A comprehensive comparison with experiment requires that one allow for all relevant QCD physical effects, including higher twist contributions and nonperturbative effects, particularly in jet fragmentation phenomenology, as well as initial and final state interactions and other non-leading contributions. It now seems apparent that these complications are preventing a detailed, quantitative check of the theory: e.g. determinations of $\alpha_s(Q^2)$ still have uncertainties at the 50% level.[33] Some of the complications which plague present QCD tests are listed in Table II.

Thus in order to really test QCD quantitatively we will need considerable information from nonperturbative dynamics. In particular, a detailed understanding of hadronic wave functions is needed in order to analyze the shape and Q^2 behavior of structure functions, the form of fragmentation distributions in $k_\perp$ and x, the effects of initial state interactions and how they control $k_\perp$ smearing effects, the form of distribution amplitudes needed for analyzing exclusive processes, as well as

Table II

Physics Measurements	QCD Complications
$[R_{e^+e^-}/3\Sigma e_q^2 - 1]$	Needs high precision; smeared data.
Structure function evolution	Higher twist terms; heavy quarks threshold effects; EMC effect.
$e^+e^- \to Jets$	Fragmentation model dependence.
$\Upsilon \to$ hadrons (3 jets)	Poor convergence of perturbation theory.
$p\bar{p} \to \ell\bar{\ell}X$	Poor convergence of perturbation theory (k–factor); no proof of factorization beyond two loop.
$pp \to HX,\ \gamma X$	Nonperturbative smearing corrections; initial and final state interactions; higher twist terms; k-factors.
$F_{2\gamma}(x, Q^2)$	Higher order QCD corrections; relation to vector meson dominated hadronic component not well understood.
$G_M^P(Q^2)$ and other exclusive channels	Higher order corrections not known; complications from end-point region; soft-wave function background; pinch singularities in hadron–hadron scattering.

calculating most power-law suppressed higher-twist contributions. Solutions to this problem await further progress in solving the light-cone equation of state or the equivalent. In the next section we will discuss a new approach for solving the scale and scheme ambiguities of perturbative QCD expansions. The present status of the factorization problem for inclusive hadronic reactions is discussed in Section 3.

2. Perturbative Expansions in Gauge Theories[34]

One of the most serious problems confronting the quantitative interpretation of QCD is the ambiguity concerning the setting of the scale in perturbative expansions. As an example, consider the standard perturbative expansion for the e^+e^- annihilation cross section in ($\overline{MS}$ scheme)

$$\left[\frac{R_{e^+e^-}(Q^2)}{3\sum e_q^2}-1\right]=\frac{\alpha_s^{\overline{MS}}(Q^2)}{\pi}\left[1+(1.98-0.115\,n_f)\frac{\alpha_s}{\pi}+O\left(\frac{\alpha_s^2}{\pi^2}\right)+\cdots\right]. \tag{2.1}$$

were n_f is the number of light fermion flavors with $m_f^2 \ll Q^2$. Note that if one chooses a different scale $Q \to \kappa Q$ in the argument $\alpha_s^{\overline{MS}}$ then the coefficient of all subsequent terms are changed. If this were a true ambiguity of QCD then higher order perturbative coefficients are not well-defined; furthermore, there is no clue toward the convergence rate of the expansion.

Is the scale choice really arbitrary? Certainty it is not arbitrary in QED. The running coupling constant is defined as

$$\alpha(Q^2)=\frac{\alpha(Q_0^2)}{1-\alpha(Q_0^2)[\pi(Q^2)-\pi(Q_0^2]} \tag{2.2}$$

where $\pi(Q^2)$ sums the proper contributions to the vacuum polarization. In lowest order QED

$$\pi(Q^2)-\pi(Q_0^2)\simeq n_f\frac{\alpha}{3\pi}\log\frac{Q^2}{Q_0^2}\,. \tag{2.3}$$

The use of the running coupling constant simplifies the form of QED perturbative expansions. For example, the light flavor contributions to the muon anomalous moment is automatically summed when we use the form

$$a_\mu=\frac{\alpha(Q^*)}{2\pi}+0.327\ldots\frac{\alpha^2(Q^*)}{\pi^2}+\cdots \tag{2.4}$$

where the scale Q^* is chosen such that[35]

$$\alpha(Q^*)=\frac{\alpha}{1-\frac{\alpha}{\pi}\left(\frac{2}{3}\log\frac{m_\mu}{m_e}-\frac{25}{18}\right)+\cdots}\,. \tag{2.5}$$

The scale Q^* in Eq. (2.4) is in fact unique; it is defined via Eq. (2.5) in such a way as to automatically sum all vacuum polarization contributions. The form of Eq. (2.4) is invariant as one changes the overall scale (e.g. $m_\mu \to m_\tau$) as we pass each new flavor threshold, if the vacuum polarization contribution of each new flavor is included in (2.5). Note, however, that the light-by-light contribution to a_μ, which appears in order α^3/π^3 from light-flavor box graphs, is not included in $a_\mu(Q^*)$ since this contribution is not part of the photon propagator renormalization and it does not contribute as a geometric series in higher order. Furthermore, for some QED processes, e.g. orthopositronium decay

$$\Gamma_{\text{orthopositronium}\to 3\gamma} \propto \alpha^3 m_e \left[1 - 10.3\ \frac{\alpha}{\pi} + \cdots\right] \tag{2.6}$$

there are no vacuum polarization corrections to this order, so the large coefficient cannot be avoided by resetting the scale in α. In QED, the running coupling constant simply sums $\alpha(Q)$ vacuum polarization contributions; in effect there are no scale-ambiguities for setting the scale. Similarly in QCD, it must be true that the vacuum polarization due to light fermions should be summed in $\alpha_s(Q)$. In fact, as we show below, this natural requirement automatically and consistently fixes the QCD scale for the leading non-trivial order in α_s for most QCD processes of interest.

In QCD the running coupling constant satisfies

$$\alpha_s(Q^*) = \frac{\alpha_s(Q)}{\left[1 + \frac{\beta_0}{2\pi}\alpha_s(Q)\ell n\left(\frac{Q^*}{Q}\right) + \cdots\right]} \tag{2.7}$$

where $\beta_0 = 11 - 2/3 n_f$. Consider any observable $\rho(Q)$ which has a perturbative expansion at large momentum transfer Q. For definiteness we choose the $\overline{MS}$ renormalization scheme to define the renormalization procedure, and adopt the canonical form,

$$\rho(Q) = \frac{\alpha_{\overline{MS}}(Q)}{\pi}\left[1 + \frac{\alpha_{\overline{MS}}}{\pi}(A_{vp} n_f + B) + \cdots\right].$$

The second order coefficient can also be written as $-\frac{3}{2}\beta_0 A_{vp} + \left(\frac{33}{2}\,A_{vp} + B\right)$. The requirement that the fermion vacuum polarization contribution is absorbed into the running coupling constant plus the fact that $\alpha(Q)$ is a function of n_f through β_0 then uniquely sets the scale of the leading order coefficient:

$$\rho(Q) = \frac{\alpha_{\overline{MS}}(Q^*)}{\pi}\left[1 + \frac{\alpha_{\overline{MS}}}{\pi}C_1 + \cdots\right] \tag{2.8}$$

where $Q^* \equiv Q\, e^{3A_{vp}}$ and $C_1 = B + \frac{33}{2}A_{vp}$. For example, from Eq. (2.1) we have

$$\rho_R(Q) \equiv \left(\frac{R_{e^+e^-}(Q^2)}{3\Sigma e_q^2} - 1\right) = \frac{\alpha_{\overline{MS}}(0.71\,Q)}{\pi}\left[1 + 0.08\ \frac{\alpha_{\overline{MS}}}{\pi} + \cdots\right].$$

Thus Q^* and C_1 are determined unambiguously within this renormalization scheme and are each n_f-independent. Note that the expansion is unchanged in form as one passes through a new quark threshold. Given any renormalization scheme, the above procedure automatically fixes the scale of the leading order coefficient for the non-Abelian theory. In higher orders one must carefully identify the correct $n_f\ A_{vp}$ terms; e.g. distinguish light-by-light or trigluon fermion loop contributions not associated with the definition of $\alpha_s(Q)$.

If we apply the procedure (2.8) to the QCD interaction potential between heavy quarks, then one obtains

$$V(Q) = -C_F \frac{4\pi\, \alpha_{\overline{MS}}(Q^*)}{Q^2}\left[1 - 2\frac{\alpha_{\overline{MS}}}{\pi} + \cdots\right] \tag{2.10}$$

where $Q^* = e^{-5/6}\,Q \cong 0.43Q$. Thus the effective scale Q^* in $\overline{MS}$ is $\sim 1/2$ of the "true" momentum transferred by $V(Q)$.

The results (2.9), (2.10) suggest that $R_{e^+e^-}$ or $V(Q)$ can be used to define and normalize $\alpha_s(Q)$. Such empirical definitions serve as a renormalization scheme alternative to $\overline{MS}$. For example, in principle we can define

$$\frac{\alpha_R(Q)}{\pi} \equiv \left[\frac{R_{e^+e^-}(Q)}{3\Sigma e_q^2} - 1\right] \tag{2.11}$$

as a physical definition of $\alpha_s(Q)$ analogous to the Coulomb scattering definition of α in QED. Note then that $\alpha_R(Q)$ and $\alpha_{\overline{MS}}(0.71\,Q)$ are effectively interchangeable.

A further benefit of the "automatic scale fixing procedure" is that the physical characteristics of the effective scale can be understood. For example, the evolution of the non-singlet moments is uniquely written in the form

$$\frac{\partial}{\partial \ell n\, Q^2}\, \ell n\, M_n(Q^2) = -\frac{\gamma_n^0}{8\pi}\alpha_{\overline{MS}}(Q^*)\left[1 - \frac{\alpha_{\overline{MS}}(Q_n^*)}{\pi}C_n + \cdots\right] \tag{2.12}$$

with

$$\begin{aligned} Q_2^* &= 0.48\,Q\,, \quad C_2 = 0.27 \\ Q_{10}^* &= 0.21\,Q\,, \quad C_{10} = 1.1 \end{aligned} \tag{2.13}$$

and $Q_n^* \sim Q/\sqrt{n}$ for large n. This dependence on $\sqrt{n}$ reflects the physical fact that the phase space limit on the gluon radiation causing the Q^2-evolution decreases in the large n, $x \to 1$ regime.

In the case of Υ decay, the scale-fixed form of the Lepage-Mackenzie[36] calculation is

$$\frac{\Gamma(\Upsilon \to hadrons)}{\Gamma(\Upsilon \to \mu^+\mu^-)} = \frac{10(\pi^2-9)}{81\pi e_b^2}\,\frac{\alpha_{\overline{MS}}(Q^*)}{\alpha_{QED}^2}\left(1 - \frac{\alpha_{\overline{MS}}}{\pi}(14.0 \pm 0.5) + \cdots\right) \qquad (2.14)$$

where $Q^* = 0.157 M_\Upsilon$. Thus, just as in the case for orthopositronium, a large second order coefficient is unavoidable. Other procedures which reduce or eliminate this coefficient by an ad hoc procedure are clearly incorrect if they are invalid in QED.

As we have discussed in Section 1, there is presently no really reliable method for determining $\alpha_s(Q)$ to better than $\pm$ 50% accuracy. The $\Gamma(\Upsilon \to 3g)/\Gamma(\Upsilon \to e^+e^-)$ ratio appears to be unreliable in view of the poor convergence of the perturbative expansion. A somewhat more hopeful process is the direct γ branching ratio:

$$\frac{\Gamma(\Upsilon \to \gamma_D + hadrons)}{\Gamma(\Upsilon \to hadrons)} = \frac{36\, e_b^2}{5}\,\frac{\alpha_{QED}}{\alpha_{\overline{MS}}(Q^*)}\left[1 + \frac{\alpha_{\overline{MS}}(Q^*)}{\pi}(2.2 \pm 0.6) + \cdots\right] \qquad (2.15)$$

where again $Q^* = 0.157\ M_\Upsilon$.

The automatic scale setting procedure should have general utility for evaluating the natural scale in a whole range of physical processes. In the case of some reactions such as hadron production $H_A H_B \to H_C X$ at large p_T each parton structure function has its own scale $\sim Q^2(1-x_i)$. In addition each hard scattering amplitude has a scale determined by corresponding fermion loop vacuum polarization contributions.

3. Factorization for High Momentum Transfer Inclusive Reactions[37]

One of the most important problems in perturbative QCD in the last two years has been to understand the validity of the standard factorization ansatz for hadron-hadron induced inclusive reactions. Although factorization is an implicit property of parton models, the existence of diagrams with color exchanging initial state interactions at the leading twist level has made the general proof of factorization in QCD highly problematical.

To see the main difficulties from a physical perspective, consider the usual form assumed for massive lepton pair production [see Fig. 1(a)]

$$\frac{d\sigma}{dx_1 dx_2}(H_A H_B \to \ell\bar{\ell}X) = \\ \times \frac{1}{3}\,\frac{4\pi\alpha^2}{3Q^2}\sum_i Q_i^2\left[q_A^{(1)}(x_i, Q)\,\bar{q}_B^{(2)}(x_2, Q) + (1 \to 2)\right] \qquad (3.1)$$

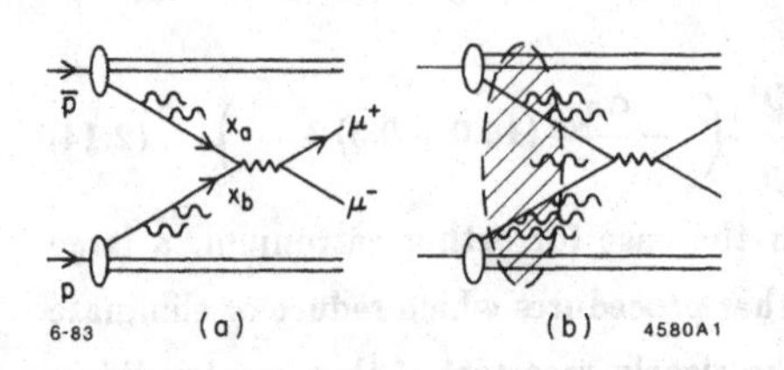

Fig. 1. (a) Gluon emission associated with QCD evolution of structure functions for the Drell-Yan process, $p\bar{p} \to \mu^+\mu^- X$. (b) Gluon emission associated with initial state interactions for the Drell-Yan process. The shaded area represents elastic and inelastic scattering of the incident quarks.

The factorization ansatz identifies the Q^2-evolved quark distributions q_A and $\bar{q}_A$ with those measured in deep inelastic lepton scattering on H_A and H_B. However, for very long targets the initial-state hadronic interactions occurring before the $q\bar{q} \to \ell\bar{\ell}$ annihilation certainly lead to induced radiation and energy loss, secondary beam production, transverse momentum fluctuations, etc. – i.e.: a profound modification of the incoming hadronic state [see Fig. 1(b)]. Since the structure functions associated with deep inelastic neutrino scattering are essentially additive in quark number even for macroscopic targets, Eq. (3.1) can obviously not be valid in general. At the least, an explicit condition related to target length must occur. The original proofs of factorization in QCD for the Drell-Yan process ignored the (Glauber) singularities associated with initial state interactions and thus had no length condition.

The potential problem and complications associated with "wee parton" exchange in the initial state were first mentioned by Drell and Yan[38] in their original work. Collins and Soper[39] have noted that proofs of factorization for hadron pair production in $e^+e^- \to H_A H_B X$ could not be really extended to $H_A H_B \to \ell\bar{\ell}X$ because of the complications of initial state effects. Possible complications associated with nonperturbative interaction effects were also discussed by Ellis *et al.*[40] More recently Bodwin, Lepage, and I[6,41] considered the effects initial state interactions as given by perturbative QCD and showed that specific graphs such as those in Fig. 2 lead to color exchange correlations as well as $k_\perp$ fluctuations. We also showed that induced hard collinear gluon radiation is indeed suppressed for incident energies large compared to a scale proportional to the length of the target. More recently, the question of the existence of color correlations on perturbative QCD has now been addressed systematically to two loop order by Lindsay *et al.*[7] and by Bodwin *et al.*[6] One finds that because of unitarity and local gauge invariance to two loop order the factorization theorem for $d\sigma/dQ^2\,dx_L$ is correct when applied at high energies to color singlet incident hadrons; more general proofs beyond two loop order await further work. We discuss the progress in this area at the end of this section.

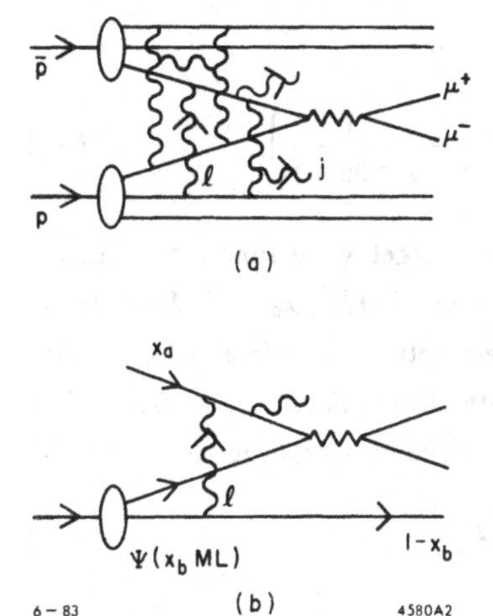

Fig. 2. (a) Representation of initial state interactions in perturbative QCD.
(b) Simplest example of induced radiation by initial state interactions in $q\pi \to \ell\bar{\ell}X$. Two different physical radiation process are included in this Feynman amplitude depending on whether the intermediate state before or after the gluon emission is on-shell. The two bremsstrahlung processes destructively interfere at energies large compared to a scale proportional to the target length L.

In addition to the above initial state interaction there are additional potential infrared problems in the non-Abelian theory associated with the breakdown of the usual Block-Nordsiek cancellation for soft gluon radiation. The work of Ref. 42 showed that any observable effect is suppressed by powers of s at high energies, again to at least two loop order.[43]

In addition to these problems the high transverse momentum virtual gluon corrections to the $q\bar{q} \to \ell\bar{\ell}$ vertex lead to relatively large radiative corrections of order $[1 + \pi^2 C_F(\alpha_s(Q^2)/\pi)]$.[44] It is usually assumed that such corrections exponentiate. As in the case of the $\Upsilon \to 3g$ problem, these corrections spoil the convergence of the perturbation theory and cannot be eliminated by choice of scale or scheme.

The remarkable feature of the QCD calculation is the fact that factorization is not destroyed by induced radiation in the target for high energy beams. This can be understood in terms of the "formation zone" principle of Landau and Pomeranchuk:[45] a system does not alter its state for times short compared to its natural scale in its rest frame. More specifically for QCD (in the Glauber/classical scattering region), consider the diagrams for induced radiation for quark-pion scattering shown in Fig. 2(b). Here $\ell^{\pm} = \ell^0 \pm \ell^3$, $y = \ell^+/p_B^+$, $x_a = p_a^-/p_A^-$ are the usual light-cone variables. The Feynman propagators of the line before and after radiation are proportional to $y - y_1 + i\epsilon$ and $y - y_2 + i\epsilon$, where the difference of the pole contribution is $y_1 - y_2 = \mathcal{M}^2/x_a s$, and $\mathcal{M}^2$ is the mass of the quark-gluon pair after bremsstrahlung. Using partial fractions, the gluon emission amplitude is then proportional to

$$\int_0^1 dy\, \psi[(x_b - y)M_n L]\left[\frac{1}{y - y_1 + i\epsilon} - \frac{1}{y - y_2 + i\epsilon}\right] \tag{3.2}$$

where we have indicated the dependence on the target wave function on target length. The two poles thus cancel in the amplitude if $(M^2/x_a s)\, M_N L \ll 1$; i.e. the radiation from the two Glauber processes destructively interfere and cancel for quark energies large compared to the target length. If we take $M^2 \sim \mu^2$ finite, then since $Q^2 = x_a x_b s$, the condition for no induced radiation translates to

$$Q^2 \gg x_b M_N L \mu^2 . \tag{3.3}$$

Taking $\mu^2 \sim 0.1\ GeV^2$, this is $Q^2 \gg x_b (0.25\, GeV^2)\, A^{2/3}$; thus one requires $Q^2 \gg x_b(10\ GeV^2)$ to eliminate induced radiation in Uranium targets.

Equation (3.3) is a new necessary condition for QCD factorization; it is also a prediction that a new type of nuclear shadowing occurs for low Q^2 lepton-pair production. If this condition is not met then the cancellations found in Ref. 7, for example, fail. The same length condition affects all sources of hard collinear radiation induced by initial or final state interactions of the hadrons or quarks in a nucleus; i.e., effectively hard collinear radiation occurs outside the target at high energies. In particular, the fast hadron production from jet fragmentation in $\ell p \to \ell H X$ occurs outside the target. In the case of very long or macroscopic targets the induced radiation destroys any semblance of factorization.

Although induced hard collinear radiation cancels at high energies, the basic processes of $k_\perp$ fluctuations from elastic collisions and induced central radiation [e.g. Fig. 2(a) with $j_z \sim m/\sqrt{s}$ in the CM] do remain. One expects that the main effects of initial state interactions can be represented by an eikonal picture where the hadronic wave functions are modified by a phase in impact space (see Fig. 3):

$$\psi_A(x_s, \vec{z}_{a\perp})\psi_B(x_b, \vec{z}_{\perp b}) \to \psi_A(x_a, \vec{z}_{a\perp})\psi_B(x_a, \vec{z}_{b\perp})U(\vec{z}_{\perp i}) . \tag{3.4}$$

Here

$$U(\vec{z}_{\perp b}) = P_T \exp\left\{-i \int_{-\infty}^{0} d\tau\, H_I(z_\perp, \tau)\right\} \tag{3.5}$$

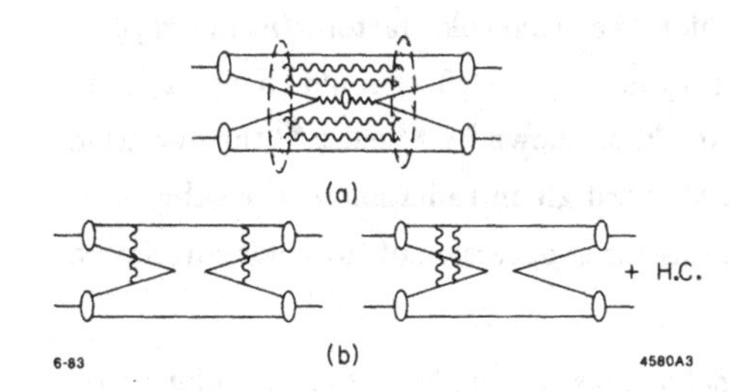

Fig. 3. (a) Representation of initial state interactions in the Drell-Yan cross section $d\sigma/dQ^2dx$. (b) Example of two-loops initial state interactions which cancel by unitarity in an Abelian gauge theory. In QCD these two contributions have different color factors.

includes elastic and soft inelastic collisions which occur up to the time $\tau = 0$ of the $q\bar{q}$ annihilation. The eikonal leads to an increased transverse smearing of the lepton pair and increased associated radiation in the central region proportional to the number of collisions ($A^{1/3}$) of the quark in the target. For a nucleus we thus predict

$$\Delta\langle Q_\perp^2\rangle \propto A^{1/3}\,, \qquad \Delta\frac{dN}{dy} \propto A^{1/3} \tag{3.6}$$

In the case of an Abelian gauge theory the integrated cross section

$$\int \frac{d\sigma}{dQ^2\,dx_L\,d^2Q_\perp} = \frac{d\sigma}{dQ^2 dx_L} \tag{3.7}$$

in unchanged because of unitarity, $U^\dagger(z_\perp)U(z_\perp) = 1$. See Fig. 3(b). Thus for an Abelian theory the increased production at large $Q_\perp$ from initial state interactions must be compensated by a depletion at low $Q_\perp$.

In general, initial state interactions will have a strong modifying effect on all hadron-hadron cross sections which produce particles at large transverse momentum simply because of the $k_\perp$ smearing of very rapidly falling distributions. The initial state exchange interactions combine with the quark and gluon $k_\perp$ distributions intrinsic to the hadron wave functions as well as that induced by the radiation associated with QCD evolution to yield the total $k_\perp$ smearing effect. The unitarity structure of the initial state eikonal interactions provides a finite theory of $k_\perp$ fluctuations even when the hard scattering amplitude is singular at zero momentum transfer.

In a non-Abelian theory the eikonal unitary matrix $U(z_\perp)$ associated with the initial state interactions is a path-color-ordered exponential integrated over the paths of the incident constituents. Since U is a color matrix it would not be expected to commute with the Drell-Yan $q\bar{q} \to \ell\bar{\ell}$ matrix element

$$U^\dagger \mathcal{M}_{DY}^\dagger \mathcal{M}_{DY} U \neq \mathcal{M}_{DY}^\dagger \mathcal{M}_{DY}\,.$$

Thus unless U is effectively diagonal in color, the usual color factor $1/n_c$ in $d\sigma(q\,\bar{q} \to \ell\bar{\ell})$ would be expected to be modified. In principle, this effect could change $1/n_c$ to n_c or 0 without violating unitarity, although, as shown by Mueller,[46] the deviation from $1/n_c$ will be dynamically suppressed; hard gluon radiation at the subprocess vertex leads to asymptotic Sudakov form factor suppression of the color correlation effect.

Despite these general possibilities, it has now been shown that the color correlation effect actually cancels in QCD at least through two loop order, although it is present in individual diagrams. The cancellation in two loops was first demonstrated in perturbation theory by Lindsay, Ross, and Sachrajda[7] for scalar quark QCD interactions in both Feynman and light-cone gauge, and was subsequently confirmed in Feynman gauge by Bodwin *et al.*[6] A detailed physical explanation of the two-loop cancellation is not known; it seems to be a consequence of both causality at high energies and local gauge invariance, although neither by itself is sufficient. We also find that the cancellation breaks down at low energies or for long targets when condition (3.3) is not satisfied. It also fails in the case of spontaneous broken gauge theories with heavy gauge boson exchange because the trigluon graph is suppressed.

An example of the nature of the color correlation cancellations is shown in Fig. 4 for $\pi\pi \to \ell\bar{\ell}X$. The diagrams shown are a gauge-invariant distinct class which have a non-trivial non-Abelian color factor and involve interactions with each of the incident spectators. The generality of the pion wave function precludes shifting of the transverse momentum interactions to other graphs. The various virtual two-gluon exchange amplitudes interfering with the zero gluon exchange amplitude each produces a $C_F C_A$ contribution which cancel in the sum. On the other hand, the imaginary part of the virtual graphs gives a non-zero contribution which potentially could lead to a color correlation at four loops. However, we find that even the imaginary part is cancelled when one includes the real emission diagrams of Figs. 4(d) and 4(e). Explicitly the sum of all the virtual and real emission amplitudes is proportional to

$$\left(C_F^2 - \frac{C_F C_A}{2}\right) \frac{2\vec{\ell}_{1\perp} \cdot \vec{\ell}_{1\perp}}{\ell_{1\perp}^2 \ell_{2\perp}^2} \frac{1}{\ell_1^+ + i\epsilon} \frac{1}{\ell_2^- + i\epsilon} \times \frac{1}{\ell_1^+ \ell_2^- - (\vec{\ell}_{1\perp} + \vec{\ell}_{2\perp})^2 - i\mathcal{E}(-\ell_1^+)} \tag{3.9}$$

The integration over ℓ_2^- then leads to zero contributions for the leading power behavior.

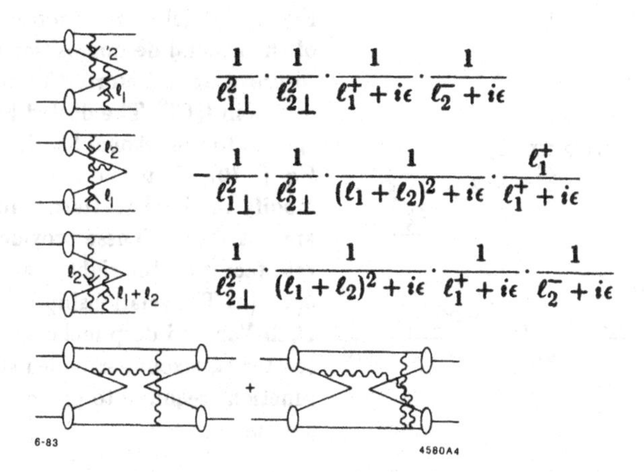

Fig. 4. Representative active spectator initial state interactions for $\pi\pi \rightarrow \ell\bar{\ell}X$ in QCD involving $C_F C_A$ evaluated in Feynman gauge. The real part of the two loop contributions represented by (a), (b), (c) (including mirror diagrams) vanishes at high energies. The imaginary parts cancel against the gluon emission contribution represented in (d) and (e).

More generally, the proof of factorization of the Drell-Yan cross section can be divided into two distinct steps, as indicated in Fig. 5. The first step is to prove that every contribution to initial state interactions in hadron-hadron scattering can be written as the convolution of two "eikonal-extended" structure functions as indicated in Fig. 5(a). This is the "weak-factorization" ansatz proposed by Collins, Soper, and Sterman[47] where each structure function has a eikonal factor attached which includes all of the elastic and inelastic initial state interactions of the corresponding incident annihilating quark or anti-quark. Explicitly, the eikonal-extended structure function of the target system A is defined as[47]

$$\begin{aligned} P_{q/A}(x,k_\perp) &= \frac{1}{2(2\pi)^3}\int dy^- \int d^2y_\perp \, e^{i(xP_A^+ y^- - \vec{k}_\perp\cdot\vec{y}_\perp)} \\ &\quad \times \langle A|\,\bar{\psi}_{DY}(0,y^-,\vec{y}_\perp)\gamma^+\psi_{DY}(0,0,\vec{0}_\perp)|A\rangle \end{aligned} \tag{3.10}$$

where

$$\Psi_{DY}(y^\nu) = P\,\exp - ig\int_{-\infty}^{0} d\lambda\, n\cdot A(y^\nu + \lambda n^\nu)\psi(y^\nu)$$

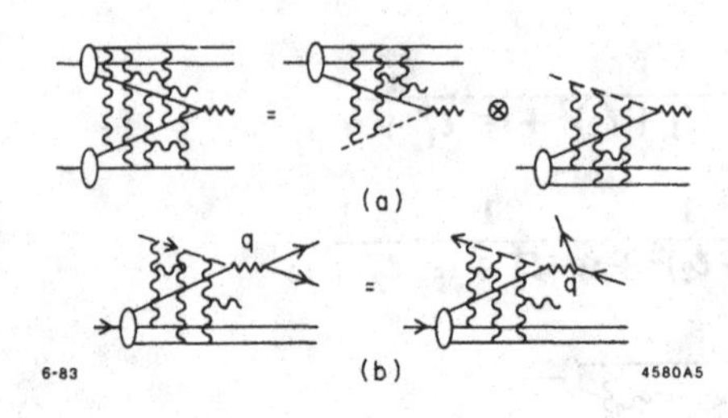

Fig. 5. (a) Schematic representation of the general decomposition required to prove weak factorization to general orders in QCD. The dotted line corresponds to the eikonal line integral of Eq. (3.10). Vertex corrections which modify the hard scattering amplitude are not shown. These provide a separate factor on the right hand side of 5(a). (b) The relationship between Drell-Yan and deep inelastic lepton scattering eikonal-extended structure functions required to prove factorization.

and n^μ is chosen such that $n \cdot \ell = 2\ell^3$ in the center-of-mass frame. The path-ordered exponential contains all of the interactions of the eikonal anti-quark line with the color gauge field along the incident $\hat{z}$ direction up to the point of annihilation.

Recently, we have in fact verified[6] that the weak factorization ansatz is correct through two loops in perturbation theory for $\mathcal{M}(A+B \to \ell\bar{\ell}X)$ despite the complicated color-topological structure of the contributing diagrams. The proof relies on splitting each Feynman amplitude into separate structure functions using identities of the form

$$\frac{1}{A\ell^+ + i\epsilon}\,\frac{1}{-B\ell^- + i\epsilon} = \left(\frac{1}{A\ell^+ + i\epsilon}\,\frac{1}{B} + \frac{1}{-B\ell^- + i\epsilon}\,\frac{1}{A}\right)\frac{1}{\ell^+ - \ell^- + i\epsilon} \tag{3.11}$$

and then analytically continuing each contribution out of the Glauber regime to either large ℓ^- or large ℓ^+, corresponding to exchange gluons collinear with the beam or target, respectively. Finally, the use of collinear Ward identities allows one to organize gauge-related diagrams into the desired weak factorization form. We are continuing efforts to try to extend the proof beyond two loop order in QCD.

The second step required to prove factorization is to show that the structure function (3.10) is actually identical to the corresponding eikonal-extended structure function for deep inelastic-lepton-hadron scattering which includes a post-factor for the final state interactions of the struck quark [see Fig. 5(b)]. This becomes intuitively obvious when one examines moments of the two structure functions. These moments differ only by terms proportional to powers of integral $\int_{-\infty}^{0} dz\, E_z(z)$, where E_z is the longitudinal component of the chromo-electric field along the eikonal line. In the center of momentum frame the hadron has ultrarelativistic momentum along the z axis, and consequently the Lorentz transformed longitudinal electric fields in the hadron are vanishing small. Thus all the moments and therefore the structure functions themselves become identical as $x \to \infty$. Physically, the effective equality of the structure functions implies that the color fluctuations generated by initial and final interactions at high energies in massive lepton pair production and deep inelastic lepton scattering are basically equivalent.[48]

At this point there is no convincing counterexample to standard QCD factorization for hadron-induced large momentum transfer reactions; on the other hand there is no proof beyond two loop order for non-Abelian theories. Clearly if factorization is a general feature of gauge theories, then it is a profound feature which demands explanation in fundamental terms.[49] In any event, the initial state interactions lead to new physical phenomena for the $Q_\perp$ distributions, especially the nuclear number dependence as well as predictions for associated particle production. Furthermore, color correlations and breakdown of factorization explicitly occur for power-law suppressed contributions which are sensitive to the length scale of the target. Such effects should be measureable for heavy nuclear targets at moderate Q^2.

A detailed discussion including applications to other processes will be discussed in Ref. 50. Here we will only mention the following.

1. If color correlations exist at higher loop order, they will be inevitably suppressed by Sudakov form factors.[46]
2. Exclusive processes factorization is unaffected at the leading power law by initial or final state interactions.[6] Physically, the hard scattering exclusive amplitudes involve only that part of the hadron wave function which corresponds to valence quarks at separation $b_\perp \sim O(1/Q)$. Strong color cancellations eliminate strong interactions of these "small" color singlet configurations. This prediction can be tested experimentally by checking whether quasi-elastic large momentum transfer exclusive reactions occur in nuclei without target-induced $k_\perp$ smearing or radiation.[51] The color transparency of small color singlet systems can also be tested by observing interactions of heavy quark anti-quark states, and also in a very interesting manner using diffractive jet phenomena of hadrons in nuclei.[52]
3. It is important to note that aside from power-law suppressed contributions, "direct" photons or hadrons do not suffer initial or final state interactions.[6] For example direct mesons produced at large transverse momentum in the subprocess $gq \to M_D q$ or direct baryons produced by the subprocess $qq \to$

B_{Dq}, have suppressed color hadronic interactions to leading order in $1/p_T^2$. Thus one can use direct photon reactions, photoproduction, Compton scattering, and direct hadron interactions especially the A-dependence of the cross sections to eliminate and effectively isolate the effects of initial and final state interactions.

REFERENCES

1. Reviews of perturbative QCD are given in A. Basetto, M. Ciafaloni, G. Marchesini, SNS 3/1983 (to be published in Physics Reports); A. H. Mueller, Phys. Rep. 73C, 237 (1981), and in Perturbative Quantum Chromodynamics, Tallasassee (1981); A. J. Buras, Rev. Mod. Phys. 52, 199 (1980). S. J. Brodsky, T. Huang, G. P. Lepage, SLAC-PUB-2868, published in Springer Vol. 100, "Quarks and Nuclear Forces", edited by D. Fries and B. Zeitnitz (1982). For a recent discussion of the multiplicity and hadron distributions predicted in QCD, see A. Mueller, CU-TP-263 (1983).
2. G. P. Lepage and S. J. Brodsky, Phys. Rev. D22, 2157 (1980).
3. S. J. Brodsky, G. P. Lepage and P. Mackenzie, Phys. Rev. D28, 228 (1983).
4. S. J. Brodsky and G. P. Lepage, SLAC-PUB-2294, published in "Quantum Chromodynamics," edited by Wm. Frazer and F. Henyey (AIP, 1979), Phys. Lett. 87B, 359 (1979); S. J. Brodsky, Y. Frishman, G. P. Lepage and C. Sachrajda, Phys. Lett. 91B, 239 (1980). See also A. V. Efremov and A. V. Radyushkin, Rev. Nuovo Cimento 3, 1 (1980), Phys. Lett. 94B, 245 (1980); A. Duncan and A. Mueller, Phys. Rev. D21, 1636 (1980), Phys. Lett. 90B, 159 (1980); G. R. Farrar and D. R. Jackson, Phys. Rev. Lett. 43, 246 (1979); V. L. Chernyak and A. R. Vhitnishii, JETP Lett. 25, 11 (1977); G. Parisi, Phys. Lett. 43, 246 (1979); M. K. Chase, Nucl. Phys. B167, 125 (1980). V. N. Baier and A. G. Grozin, Nucl. Phys. B192, 476 (1981).
5. E. Braaten, University of Florida preprint UFTP-82-11 (1982); F. del Aguila and M. K. Chase, Nucl. Phys. B193, 517 (1981). Higher order corrections to the distribution amplitude and its evolution equation are discussed by S. J. Brodsky, Y. Frishman and G. P. Lepage, in preparation.
6. G. T. Bodwin, S. J. Brodsky, and G. P. Lepage, Phys. Rev. Lett. 47, 1799 (1983) and SLAC-PUB-2966, published in the proceedings of the XIII International Symposium on Multiparticle Dynamics, Volendam, The Netherlands

(1982), SLAC-PUB-2860, to be published in the Proceedings of the Banff Summer School on Particles and Fields, 1981.

7. W. W. Lindsay, D. A. Ross, and C. T. Sachrajda, Phys. Lett. 117B, 105 (1982), Nucl. Phys. B, 214 (1983), and Southampton preprint 82/83-4 (1983).

8. J. D. Bjorken, Phys. Rev. D8, 4093 (1973); D. Sivers, R. Blankenbecler, and S. J. Brodsky, Phys. Rep. 23C, 1 (1976). S. M. Berman, J. D. Bjorken, J. B. Kogut, Phys. Rev. D4, 3388 (1971).

9. A review of higher twist contributions is given by S. J. Brodsky, E. L. Berger, and G. P. Lepage, Proceedings of the Drell-Yan Workshop, FNAL (1982). See also R. K. Ellis, W. Furmanski, and R. Petronzio, CERN-TH-3301 (1982) and R. Blankenbecler, S. J. Brodsky, and J. F. Gunion, Phys. Rev. D18, 900 (1978). See also the recent calculation of R. Blankenbecler, J. F. Gunion and P. Nason, SLAC-PUB-3142/UCD 83-2 (1983) who find large $(1-x)^2/Q^2$ contributions to the proton structure function. This supports the phenomenological structure for scale breaking discussed in L. F. Abbott, W. B. Atwood and R. M. Barnett, Phys. Rev. D22, 582 (1980) and I. A. Schmidt and R. Blankenbecler, Phys. Rev. D16, 1318 (1977).

10. H. D. Politzer, Nucl. Phys. B172, 349 (1980) and CALT-68-789 (1980), published in Copenhagen QCD, 934 (1980). S. J. Brodsky, R. R. Horgan, and W. E. Caswell, Phys. Rev. D18, 2415 (1978).

11. S. J. Brodsky and G. R. Farrar, Phys. Rev. Lett. 31, 1153 (1973); Phys. Rev. D11, 1309 (1975). V. A. Matveev, R. M. Muradyan and A. J. Tavkheldize, Lett. Nuovo Cimento 7, 719 (1973).

12. N. Paver and D. Treleani, Nuovo Cimento A70, 215 (1982); H. D. Politzer, Ref. 10; S. J. Brodsky and J. F. Gunion (unpublished). B. Humbert, CERN-TH-3620 (1983).

13. S. J. Brodsky, Y. Frishman, G. P. Lepage, and C. Sachrajda, Phys. Lett. 91B, 239 (1980).

14. See Refs. 2, 4 and S. J. Brodsky and G. P. Lepage, Phys. Rev. Lett. 43, 545, 1625(E) (1979); S. J. Brodsky, G. P. Lepage and S.A.A. Zaidi, Phys. Rev. D23, 1152 (1981). S. J. Brodsky and G. P. Lepage, Phys. Rev. D24, 2848 (1981).

15. P. V. Landshoff, Phys. Rev. D10, 1024 (1974). A. Mueller, Phys. Lett. 108B, 355 (1982).

16. S. J. Brodsky and G. P. Lepage, Phys. Rev. D24, 1808 (1981); P. Damgaard, Cornell preprint CLNS 81/519 (1981).

17. R. D. Field, R. Gupta, S. Otto, L. Chang, Nucl. Phys. B186, 429 (1981). F. M. Dittes and A. V. Radyushkin, Dubna preprint JINR-12280-C88 (1980).

18. S. J. Brodsky, T. Huang and G. P. Lepage, SLAC-PUB-2540 (1980); T. Huang, SLAC-PUB-2580 (1980), published in the Proceedings of the XXth International Conference on High Energy Physics, Madison, Wisconsin (1980), and in preparation; S. J. Brodsky and S. D. Drell, Phys. Rev. D22, 2236 (1981).

19. E. L. Berger and S. J. Brodsky, Phys. Rev. Lett. 42, 940 (1979). See also B. Pire and J. Ralston, in Proceedings of the Drell-Yan Workshop, FNAL, 1983.

20. E. L. Berger and S. J. Brodsky, Phys. Rev. D24, 2428 (1981).

21. J. A. Bagger and J. F. Gunion, Phys. Rev. D25, 2287 (1982), and UCD-82/1 (1983). S. J. Brodsky and J. Hiller, SLAC-PUB-3047, to be published in Phys. Rev. C.

22. The rules for light-cone perturbation theory are summarized in Ref. 2, and G. P. Lepage, S. J. Brodsky, T. Huang, and P. Mackenzie, CLNS-82/522, to appear in the Proceedings of the 1981 Banff Summer Institute on Particles and Fields. Here $[d^2k_\perp] \equiv 16\pi^3\delta\left(\sum_{i=1}^{2} k_{\perp i}\right)\prod_{j=1}^{n}(d^2k_{\perp j}/16\pi^3)$ and $[dx] = \delta\left(1-\sum_{i=1}^{n} x_i\right)\prod_{j=1}^{n} dx_j$.

23. S. J. Brodsky, G. P. Lepage, and T. Huang, Ref. 1, and V. A. Karmanov, Nucl. Phys. A362, 331 (1982).

24. See, e.g., B. L. Ioffe and A. V. Smilga, Nucl. Phys. B216, 273 (1983) and Phys. Lett. 114B, 353 (1982), and references therein.

25. See, e.g., M. S. Witherall, in Proceedings of Experimental Meson Spectroscopy — 1980, and C. Peterson in Proceedings of the 13th International Symposium on Multiparticle Dynamics, Volendam, Netherlands (1982), D. DiBitonto, Harvard Thesis RX-900 (1979).

26. S. J. Brodsky, P. Hoyer, C. Peterson and N. Sakai, Phys. Lett. 93B, 451 (1980); S. J. Brodsky, C. Peterson and N. Sakai, Phys. Rev. D23, 11 (1981); S. J. Brodsky and C. Peterson, SLAC-PUB-2888 (1982), and Proceedings of the Topical Conference on Forward Collider Physics, Madison, Wisconsin

(1981); C. Peterson, 13th International Symposium on Multiparticle Dynamics, Volendam, Netherlands (1982). R. V. Gavai and D. P. Roy, Z. Phys. C15, 29 (1982).

27. J. W. Cronin *et al.*, Phys. Rev. Lett. 31, 1426 (1973), Phys. Lett. 38, 115 (1972).

28. See, e.g., T. Akesson *et al.*, Phys. Lett. 123B, 367 (1983), E. Anassontzis *et al.*, Zeit. Phys. C13, 277 (1982).

29. J. J. Aubert *et al*, Phys. Lett. 123B, 123 (1983); A. Bodek *et al.*, SLAC-PUB-3089 (1983).

30. V. I. Zakharov and N. N. Nikolaev, Sov. J. Nucl. Phys. 21, 227 (1975). J. D. Bjorken, private communication.

31. J. Szwed, Cracow preprint 1/83 (1982). A. Bialas and J. Szwed, presented at the 1983 Zakopane Summer School. C. H. Llewellyn Smith, Oxford preprint 18/83 (1983); A. W. Thomas, CERN TH-3552 (1983); M. Ericson and A. W. Thomas, CERN Report 3553 (1983); E. L. Berger, F. Coester and R. B. Wiringa, ANL-HEP-PR-83-24 (1983).

32. H. J. Pirner and J. P. Vary, Phys. Rev. Lett. 46, 1376 (1981); R. Jaffe, Phys. Rev. Lett. 50, 228 (1983). L. S. Celenza and C. Shakin, Brooklyn College preprint BCINT-82/111/117 (1982). M. Staszel, J. Roznek, G. Wilk, Warsaw preprint IFT19/83 (1983). F. E. Close, R. B. Robert, G. G. Ross, Rutherford preprint RL-83-051 (1983). O. Nachtmann and J. H. Pirner, Heidelberg preprint HD-THE8-83-8 (fi983).

33. For a recent review, see R. Hollebeek and S. Ellis to be published in the Proceedings of the 1983 SLAC Summer Institute.

34. The work in this section is based on Ref. 3. See also W. Celmaster and R. J. Gonsalves, Phys. Rev. D20, 1420 (1979), P. M. Stevenson, Phys. Rev. D23, 2916 (1981) for further discussions regarding choice of renormalization scheme. The dependence on scheme for the procedure of Ref. 3 is discussed by W. Celmaster and P. M. Stevenson, Phys. Lett. 125B, 493 (1983).

35. This procedure for QED is equivalent to mass-singularity analyses and renormalization group methods discussed by T. Kinoshita, Nuovo Cimento 51B, 140 (1967); B. E. Lautrup and E. deRafael, Nucl. Phys. B70, 317 (1974); and M. A. Samuel, Phys. Rev. D9, 2913 (1978). A related method for summing higher-loop QCD contributions to structure function moments in deep

inelastic scattering is given in M. Moshe, Phys. Rev. Lett. 43, 1851 (1979) and A. Blumenfeld and M. Moshe (Ref. 5).

36. P. B. Mackenzie and G. P. Lepage, Phys. Rev. Lett. 47, 1244 (1981).

37. The work in this section is based on Ref. 6. A detailed report is in preparation.

38. S. D. Drell and T. M. Yan, Phys. Rev. Lett. 25, 316 (1970).

39. J. C. Collins and D. E. Soper, Proceedingso of the Moriond Workshop, Les Arce, France (1981).

40. J. E. Ellis, M. K. Gaillard, W. J. Zakrzewski, Phys. Lett. 81B, 224 (1979).

41. The light-cone gauge calculation of Ref. 6 was incomplete because of contributions from seagull diagrams, and contributions outside the Glauber regime.

42. J. Frenkel *et al.*, preprint IFUSP/P-405 (1983), and references therein.

43. For a contrary view, see H. Barnarjee *et al.*, CERN preprint TH-3544 (1983).

44. See, e.g., F. Khalafi and J. Stirling, Cambridge preprint DAMTP 83/2 (1983). For reviews, see J. Stirling in Proceedings of the XIIIth International Symposium on Multiparticle Dynamics, Volendam (1982), and A. Basetto *et al.*, Ref. 1.

45. L. Landau and I. Pomeranchuk, Dok. Akademii Nauk SSSR 92, 535 (1953), and 92, 735 (1953); L. Stodolsky, MPI-PAE/pTH 23/75 (1981). I. M. Dremin, Lebedev preprint 250 (1981).

46. A. Mueller, Phys. Lett. 108B, 355 (1982). A. Sen and G. Sterman, Fermilab-PUB-83/42 ThY (1983).

47. J. C. Collins, D. E. Soper, G. Sterman, Phys. Lett. 109B, 288 (1983). SUNY preprint ITP-SB-82-46 (1982).

48. For related work, see also J. C. Collins, D. E. Soper, G. Sterman, Phys. Lett. 126B, 275 (1983).

49. A. Mueller, Proceedings of the Drell-Yan Workshop, FNAL (1982).

50. G. T. Bodwin, S. J. Brodsky, and G. P. Lepage, in preparation.

51. Further discussion may be found in S. J. Brodsky, SLAC-PUB-2970 (1982), published in the Proceedings of the XIIIth International Symposium on Multiparticle Dynamics, and Ref. 49.

52. G. Bertsch, S. J. Brodsky, A. S. Goldhaber and J. F. Gunion, Phys. Rev. Lett. 47, 297 (1981).

"DO THE QCD CORRECTIONS DESTROY THE AMPLITUDE ZERO IN $q\bar{q} \to W\gamma$?"

N.M. Monyonko

Physics Department, University of Nairobi, P.O. Box 30197, Nairobi, Kenya

J.H. Reid

Physics Department and Theoretical Science Institute,
Simon Fraser University, Burnaby, B.C., Canada V5A 1S6

ABSTRACT

We report some preliminary results from a calculation of order α_s QCD corrections to the amplitude zero in $q\bar{q} \to W\gamma$.

It was observed by Samuel, Brown, Mikaelian and Sahdev [1] that processes involving quarks and a charged vector boson minimally coupled to a real photon exhibit zeros in the amplitude for certain values of the kinematic invariants. These intriguing 'amplitude zeros', as they are called, occur only if the couplings are minimal i.e., as prescribed by a Yang-Mills gauge theory. Their origin was investigated by Goebel, Halzen, Leveille and Dongpei [2]. Recently Brodsky and Brown [3] proved the following theorem:

"Let T_G be a tree graph with n external lines labeled by particle four-momenta p_i, charges Q_i and masses m_i. If M_γ is the single-photon emission amplitude which is the sum generated by making photon attachments with four-momentum q in all possible ways onto T_G, then $M_\gamma = 0$ if all the ratios $Q_i/p_i \cdot q$ are equal. The couplings must be minimal (i.e. of Yang-Mills type)."

Such zeros may be expected to disappear when higher order corrections are included. We wish to report some preliminary results of a calculation of the order α_s QCD corrections to the process $q\bar{q} \to W\gamma$, which underlies the reaction $p\bar{p} \to W\gamma$. This reaction may be observed eventually in $p\bar{p}$ experiments. A calculation has already been done of the QCD $O(\alpha_s)$ corrections to the related but simpler process $W \to q\bar{q}$ by Albert, Marciano, Wyler and Parsa [4]. They find the correction to the decay rate Γ_0 to be exactly $\alpha_s/\pi\Gamma_0$ i.e. $\approx 5\%$. One may expect a similar, rather small correction in the case of $q\bar{q}\to W\gamma$, which is encouraging because then the amplitude zero would not be entirely filled in, but would still give a sharp and measurable dip to the cross section.

Unfortunately the full calculation is very complicated. All we can report here are some preliminary results.

If we consider the process $q\bar{q}\to W\gamma$ in the lowest order then the total amplitude T_0 factorizes by virtue of the Brodsky-Brown theorem previously quoted as follows:

$$T_0 = \left(\frac{Q_i}{k\cdot q_1} - \frac{Q_j}{k\cdot q_2}\right)\tilde{T}_0 \tag{1}$$

where $p,k,q_{1,2}$ are vector boson, photon and quark four momenta and $Q_{i,j}$ are quark charges. It is convenient to work in the approximation of massless quarks, in which case there are not only infrared but also collinear or mass singularities to be considered. The former singularities must cancel between loop and bremsstrahlung terms (Bloch-Nordsieck theorem), while the latter must also cancel out in the final differential cross section (Kinoshita-Sirlin-Lee-Nauenberg theorem). Denoting the amplitude from 1 loop QCD—corrected graphs by T_2^{loop} and the contribution from the gluon bremsstrahlung graphs by T_1^{brems}, we see that the cross section is proportional to:

$$\text{cross section} \propto |T_0|^2 + T_0 T_2^{*loop} + T_0^* T_2^{loop} + |T_1^{brems}|^2$$
$$(\text{to order } \alpha_s) \quad . \tag{2}$$

We observe that *at the amplitude zero*, i.e., when $Q_i(k\cdot q_2) = Q_j(k\cdot q_1)$, T_0 vanishes thus wiping out the 1 loop contributions as well. Only the bremsstrahlung term remains, and this must now be *finite by itself*. In order to find out what QCD corrections do to the zero we shall work *at the zero*. Then we only have to evaluate the contributions from the 8 bremsstrahlung diagrams in figures 1,2,3.

First the zero amplitude T_0 may be written:

$$T_0 = \frac{-ie\, g_W}{2\sqrt{2}}\, \varepsilon_\gamma^\mu\, \varepsilon_W^\nu\, \bar{v}_j(q_2)\, M(1-\gamma_5)\, u_i(q_1) \tag{3}$$

where $M = [Q_i(k\cdot q_2) - Q_j(k\cdot q_1)]\, \frac{1}{k\cdot p}\left(\gamma_\nu \frac{1}{\not{\ell}_1}\gamma_\mu - \gamma_\mu \frac{1}{\not{\ell}_2}\gamma_\nu\right)$. Here

$q_1+q_2 = p+k$, $\ell_1 = p-q_2$, $\ell_2 = p-q_1$, $Q_{i,j}$ are quark charges, $k^2 = q_1^2 = q_2^2 = 0$, $p^2 = M^2$, $\varepsilon_{\gamma,W}$ are polarization vectors and $g_W^2/8M^2 = G_F$. In similar notation the bremsstrahlung contributions may be written

$$T_1^{brems} = \frac{-ie\; g_W\; g_s}{2\sqrt{2}}\, \varepsilon_\gamma^\mu \varepsilon_W^\nu \varepsilon_g^\rho\, \frac{\lambda_{ij}^a}{2}\, \bar{v}_j(q_2) M^{brems}(1-\gamma_5)u_i(q_1) \tag{4}$$

$$M = M_1 + M_2 + M_3 + \widetilde{M}_2 + \widetilde{M}_3 \;. \tag{5}$$

In figure 1 the gluon is radiated from an internal quark line, in figures 2,3 it is radiated from external lines

$$M_1 = Q_i\; \gamma_\nu \frac{1}{\not{\ell}_1} \gamma_\rho \frac{1}{\not{\ell}_1+\not{g}} \gamma_\mu + Q_j\; \gamma_\mu \frac{1}{\not{\ell}_2+\not{g}} \gamma_\rho \frac{1}{\not{\ell}_2} \gamma_\nu \tag{6}$$

$$M_2 = \left(\frac{Q_i\,(k\cdot q_2)-Q_j\; k\cdot(q_1-g)}{k\cdot p}\right)\left(\gamma_\nu \frac{1}{\not{\ell}_1} \gamma_\mu - \gamma_\mu \frac{1}{\not{\ell}_2+\not{g}} \gamma_\nu\right) \frac{1}{\not{q}_1-\not{g}} \gamma_\rho \tag{7}$$

$$M_3 = \left(\frac{Q_i\; k\cdot(q_2-g)-Q_j\; k\cdot q_1}{k\cdot p}\right) \gamma_\rho \frac{1}{\not{q}_2-\not{g}} \left(\gamma_\nu \frac{1}{\not{\ell}_1+\not{g}} \gamma_\mu - \gamma_\mu \frac{1}{\not{\ell}_2} \gamma_\nu\right) \tag{8}$$

$$\widetilde{M}_2 = \frac{(Q_i + Q_j)}{2k\cdot p}\left[\gamma_\nu\gamma_\mu\gamma_\rho - \frac{1}{2[q_1\cdot g + k\cdot(q_1-g)]}\, \gamma_\nu \not{\ell}_1 \gamma_\mu (\not{q}_1-\not{g})\gamma_\rho\right] \tag{9}$$

$$\widetilde{M}_3 = \left(\frac{Q_i + Q_j}{2k\cdot p}\right)\left[\gamma_\rho\gamma_\mu\gamma_\nu - \frac{1}{2[q_2\cdot g + k\cdot(q_2-q)]}\, \gamma_\rho (\not{q}_2-\not{g})\gamma_\mu \not{\ell}_2 \gamma_\nu\right] \;. \tag{10}$$

Note that although $M_{2,3}$ contain the usual infrared divergent terms in the denominator proportional to $q_1\cdot g$, $q_2\cdot g$, *at the zero* when $Q_i(k\cdot q_2) - Q_j(k\cdot q_1) = 0$, there is also the compensating term $k\cdot g$ in the *numerator*. This is the mechanism by which the infrared divergence is eliminated *without* the help of loop diagrams. The terms M_1, $\widetilde{M}_2$ and $\widetilde{M}_3$ are always infrared convergent, whether one is at the zero or away from it.

The total matrix element squared (spin and color summed), is a very complicated sum of hundreds of terms. We shall just look at the properties of the squared terms $|T_1|^2$, $|T_2|^2$ and $|T_3|^2$, leaving aside the more complicated cross terms. We find:

$$|T_1|^2 = (A_1 + B_1)\, \frac{e^2\; g_W{}^2\; g_s{}^2}{12} \quad \text{where} \quad g_s{}^2/4\pi = \alpha_s \tag{11}$$

$$A_1 = \frac{256Q_iQ_j\ q_1\cdot q_2\ \ell_1\cdot\ell_2(\ell_1+g)\cdot(\ell_2+g)}{\ell_1^2\ \ell_2^2(\ell_1+g)^2\ (\ell_2+g)^2}$$

$$+ 32Q_i^2\left(\frac{4q_1\cdot(\ell_1+g)q_2\cdot\ell_1\ \ell_1\cdot(\ell_1+g)}{\ell_1^4(\ell_1+g)^4} - \frac{2q_1\cdot(\ell_1+g)q_2\cdot(\ell_1+g)}{\ell_1^2(\ell_1+g)^4}\right.$$

$$\left. - \frac{2q_1\cdot\ell_1\ q_2\cdot\ell_1}{\ell_1^4(\ell_1+g)^2} + \frac{q_1\cdot q_2}{\ell_1^2(\ell_1+g)^2}\right)$$

$$+ 32Q_j^2\left(\frac{4q_2\cdot(\ell_2+g)q_1\cdot\ell_2\ \ell_2\cdot(\ell_2+g)}{\ell_2^4(\ell_2+g)^4} - \frac{2q_2\cdot(\ell_2+g)q_1\cdot(\ell_2+g)}{\ell_2^2(\ell_2+g)^4}\right.$$

$$\left. - \frac{2q_2\cdot\ell_2\ q_1\cdot\ell_2}{\ell_2^4(\ell_2+g)^2} + \frac{q_2\cdot q_1}{\ell_2^2(\ell_2+g)^2}\right) \tag{12}$$

$$B_1 = \frac{64\ Q_iQ_j(\ell_1+g)\cdot(\ell_2+g)}{M^2\ell_1^2\ell_2^2(\ell_1+g)^2(\ell_2+g)^2}\left(\begin{array}{l} 2p\cdot q_1p\cdot\ell_2q_2\cdot\ell_1+2p\cdot q_2p\cdot\ell_1q_1\cdot\ell_2 \\ -2p\cdot q_1p\cdot\ell_1q_2\cdot\ell_2-2p\cdot q_2p\cdot\ell_2q_1\cdot\ell_1 \\ -M^2\ q_1\cdot q_2\ell_1\cdot\ell_2-M^2\ q_2\cdot\ell_1q_1\cdot\ell_2 \\ +M^2q_1\cdot\ell_1q_2\cdot\ell_2 \end{array}\right)$$

$$\frac{+16\ Q_i{}^2}{M^2\ell_1^4(\ell_1+g)^4}\left(\begin{array}{l} M^2\ell_1^2[2q_1\cdot(\ell_1+g)q_2\cdot(\ell_1+g) - q_1\cdot q_2(\ell_1+g)^2] \\ -2\ell_1^2\ \ell_1\cdot q_2[2p\cdot(\ell_1+g)q_1\cdot(\ell_1+g) - p\cdot q_1(\ell_1+g)^2] \\ +2\ell_1\cdot q_2(M^2-2\ell_1\cdot q_2)[2q_1\cdot(\ell_1+g)\ell_1\cdot(\ell_1+g) \\ -q_1\cdot\ell_1(\ell_1+g)^2] \end{array}\right)$$

$$\frac{+16\ Q_j{}^2}{M^2\ell_2^4(\ell_2+g)^4}\left(\begin{array}{l} M^2\ell_2^2[2q_2\cdot(\ell_2+g)q_1\cdot(\ell_2+g) - q_2\cdot q_1(\ell_2+g)^2] \\ -2\ell_2^2\ \ell_2\cdot q_1[2p\cdot(\ell_2+g)q_2\cdot(\ell_2+g) - p\cdot q_2(\ell_2+g)^2] \\ +2\ell_2\cdot q_1(M^2-2\ell_2\cdot q_1)[2q_2\cdot(\ell_2+g)\ell_2\cdot(\ell_2+g) \\ q_2\cdot\ell_2(\ell_2+g)^2] \end{array}\right) \tag{13}$$

$$|T_2|^2 = (A_2 + B_2)\ \frac{e^2g_W{}^2g_s{}^2}{12}\left(\frac{Q_ik\cdot q_2-Q_jk\cdot(q_1-g)}{k\cdot p}\right)^2 \tag{14}$$

$$A_2 = \frac{16}{q_1\cdot g}\left(\begin{array}{l} \dfrac{+2q_2\cdot(\ell_2+g)\ell_2\cdot g}{(\ell_2+g)^4}\ \ \dfrac{-q_2\cdot g}{(\ell_2+g)^2} \\ \dfrac{+4\ell_1\cdot(\ell_2+g)q_2\cdot g}{\ell_1{}^2(\ell_2+g)^2}\ \ \dfrac{+2q_2\cdot\ell_1g\cdot\ell_1}{\ell_1{}^4}\ \ \dfrac{-q_2\cdot g}{\ell_1{}^2} \end{array}\right) \tag{15}$$

$$B_2 = \frac{-8}{M^2 q_1\cdot g}\Bigg[\left(\frac{1}{\ell_1^2}+\frac{1}{(\ell_2+g)^2}\right)(2p\cdot q_2 p\cdot g - M^2 q_2\cdot g)$$
$$+\frac{1}{\ell_1^4}(2M^2 q_2\cdot\ell_1 g\cdot\ell_1 - 4p\cdot q_2 p\cdot\ell_1\, g\cdot\ell_1)$$
$$+\frac{1}{(\ell_2+g)^4}[2M^2 g\cdot\ell_2 q_2\cdot(\ell_2+g) - 4p\cdot g\; p\cdot(\ell_2+g) q_2\cdot(\ell_2+g)]$$
$$+\frac{1}{\ell_1^2(\ell_2+g)^2}\Big[+2M^2(g\cdot q_2\ell_2\cdot\ell_1 - g\cdot\ell_1 q_2\cdot\ell_2 + q_2\cdot\ell_1 g\cdot\ell_2)$$
$$+4p\cdot q_2 g\cdot\ell_1 p\cdot\ell_2 - 4p\cdot q_2 p\cdot\ell_1 g\cdot\ell_2 - 4p\cdot q_2$$
$$g\cdot\ell_1 p\cdot g + 4p\cdot g p\cdot\ell_1 q_2\cdot\ell_2 - 4p\cdot g p\cdot\ell_1$$
$$q_2\cdot g - 4p\cdot g p\cdot\ell_2 q_2\cdot\ell_1 + 4(p\cdot g)^2 q_2\cdot\ell_1\Big]\Bigg] \qquad (16)$$

$|T_3|^2$ may be obtained from $|T_2|^2$ by the exchange of q_1,ℓ_1 with q_2,ℓ_2, and Q_i with Q_j.

In order to perform the phase space integral over the gluon momentum, we choose a center of mass frame with 4 momenta:

$$q_1+q_2 = p+k+g, \quad q_1^2 = q_2^2 = k^2 = g^2 = 0, \quad p^2 = M^2$$
$$q_1 = (E_o,\vec{q}) \quad q_2 = (E_o,-\vec{q}) \quad k = (E_1,\vec{k}) \quad g = (E_2,\vec{g}) \quad p = (E_3,\vec{p})$$
$$2E_o = W = E_1+E_2+E_3 \quad E_1 = |\vec{k}|,\ E_2 = |\vec{g}|,\ \vec{k}+\vec{g}+\vec{p} = 0$$
$$\hat{k}\cdot\hat{g} = \cos\theta_1 \quad \hat{k}\cdot\hat{q} = \cos\theta_2 \quad \hat{g}\cdot\hat{q} = \cos\theta_3$$
$$\cos\theta_3 = \sin\theta_2\,\sin\theta_1\,\cos\phi + \cos\theta_2\,\cos\theta_1 \quad .$$

Then the partial differential cross section is

$$\frac{d\sigma}{d\cos\theta_2} = \frac{1}{(2\pi)^4\; 16W^2}\iiint dE_1\, dE_2\, d\phi\; |T|^2 \qquad (17)$$

Here $\cos\theta_1$ equals $1 + \dfrac{[W^2-M^2-2W(E_1+E_2)]}{2E_1E_2}$ and the limits $\cos\theta_1 = \pm 1$ define the boundaries of the (E_1E_2) phase space: ϕ is unrestricted. The outer and inner boundaries meet at the two points $(0,\ W^2-M^2/2W)$ and $(W^2-M^2/2W,\ 0)$. These are the dangerous points for the integration since we find terms in the integrand $\propto 1/E_1{}^p$, $1/E_2{}^q$. It is necessary to examine the expressions in detail to see how well-behaved the terms are near these points. Let us look at A_2 in eq. (15). We find that

$q_2\cdot(\ell_2+g)$, $(\ell_2+g)^2$ are proportional to E_1, $q_2\cdot g$, $q_1\cdot g$, $\ell_2\cdot g$, $\ell_1\cdot g$ to E_2, and other terms are proportional to neither. From this we see that $A_2 \propto 1/E_1$, and if we are *at the zero*, we also pick up a factor of $(E_1E_2/E_1)^2 = E_2^2$. Hence the A_2 contribution to $|T_2|^2$ is $\propto (E_2^2/E_1)$. The same is true of the B_2 contribution. At first it appears one will get a log singularity at $E_1 = 0$. But the phase space boundary near E_1 introduces a convergence factor so that the final result is finite!

One has roughly

$$I = \int_0^u \frac{dE_1}{E_1} \int_{a+cE_1}^{a+bE_1} dE_2\, E_2^2 \int_0^{2\pi} d\phi$$

$$I \propto \int_0^u \frac{dE_1}{E_1}\left[(a+bE_1)^3 - (a+cE_1)^3\right] = \int_0^u dE_1\,[\text{constant} + E_1 + \ldots]$$

$$= \textit{finite} \quad .$$

By making a similar analysis on the other terms in the matrix element squared we find that:

$\lvert M_1\rvert^2 \propto 1/E_1^2 \rightarrow \ell n E_1$			
$\lvert M_{2,3}\rvert^2 \propto E_2^2/E_1 \rightarrow$ finite			
$\lvert M_1M_2\rvert \propto 1/E_1^2 \rightarrow \ell n E_1$			None of these give trouble at $E_2 = 0$
$\lvert M_1M_3\rvert \propto 1/E_1^2 \rightarrow \ell n E_1$			
$\lvert M_2M_3\rvert \propto E_2^2/E_1^2 \rightarrow \ell n E_1$			

Since we know that the final result must be finite, cancellation must occur between $|M_1|^2$ which is positive definite and divergent, and the cross terms $|M_1M_2|$, $|M_1M_3|$ and $|M_2M_3|$ which can be negative. Only $|M_2|^2$ and $|M_3|^2$ may be evaluated in isolation since these are individually finite, thanks to the amplitude zero factor. In practice since a photon energy cut-off must be used, the divergences as $E_1 \rightarrow 0$ will not cause trouble. The results of these integrations will be finite but cut-off dependent.

The necessary integrals cannot be done explicitly and must therefore be done numerically. They are exceedingly complicated and none of them have been completed yet. However, we may get a very rough idea of magnitudes by taking on average value of the integrand. This contribution to the cross section can then be compared to the zero order cross section with the amplitude zero factor *removed*. Taking a term from eq. (15), we get approximately:

$$\frac{d\sigma^{brems}}{d\cos\theta_2} \propto \frac{1}{(2\pi)^4}\frac{1}{16W^2}\left(\frac{32}{W^2}\right)\left(2\pi\frac{W^2}{4}\right)\left(\frac{e^2 g_W^2 g_s^2}{12}\right)$$

$$= \alpha\frac{\alpha_s 2M^2 G_F}{3\pi s} \quad .$$

In contrast, the zero order cross section [1] is:

$$\frac{d\sigma}{d\cos\theta_2} \propto \frac{\alpha M^2 G_F}{s} \quad .$$

Hence the ratio of these is $\sim 2\alpha_s/3\pi$, which is of the expected order.

ACKNOWLEDGMENTS

We thank Doug Beder for crucial discussions on this work and his encouragement and support. We thank Morton Laursen for pointing out an important error in the original version of this talk. We are especially grateful for the enthusiastic discussions and useful advice on this work from Mark Samuel, Kim Milton, Achin Sen, Morton Laursen and Gary Tupper.

REFERENCES

[1] R.W. Brown, D. Sahdev, K.O. Mikaelian, Phys. Rev. D20 (1979) 1164; K.O. Mikaelian, M.A. Samuel, D. Sahdev, Phys. Rev. Lett. 43 (1979) 746.

[2] C.J. Goebel, F. Halzen, J.P. Leveille, Phys. Rev. D23 (1981) 2682; Z. Dongpei, Phys. Rev. D22 (1980) 2266.

[3] S.J. Brodsky, R.W. Brown, Phys. Rev. Lett. 49 (1982) 966.

[4] D. Albert, W.J. Marciano, D. Wyler, Z. Parsa, Nucl. Phys. B166 (1980) 460.

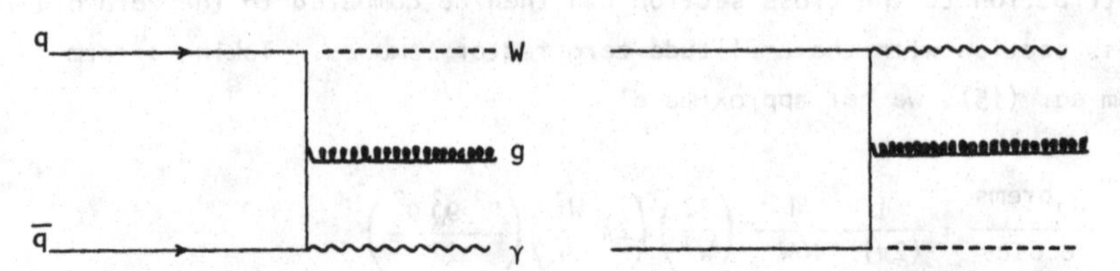

Figure 1. Gluon bremsstrahlung from an internal quark.

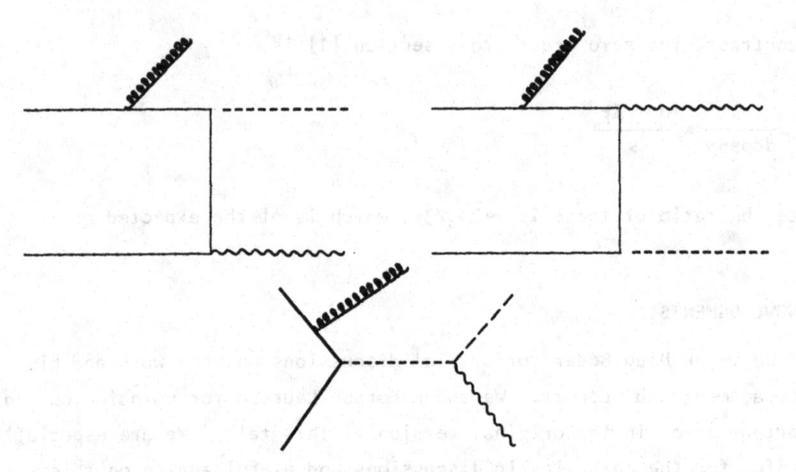

Figure 2. Gluon bremsstrahlung from an external quark.

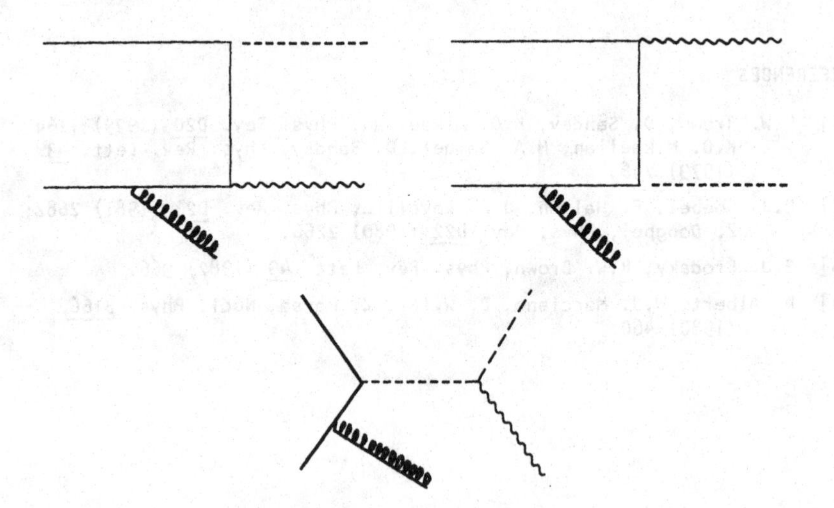

Figure 3. Gluon bremsstrahlung from an external antiquark.

THE M.I.T. BAG MODEL IN THE CONTEXT OF QCD

K. JOHNSON

Center for Theoretical Physics and Laboratory of Nuclear Science
Massachusetts Institute of Technology
Cambridge, MA 02139
USA

I. Introduction

The M.I.T. Bag Model[1,2] has been generally successful in giving a simple and quantitative account of the quark structure of hadrons. It is not the purpose of this talk to recount the successes or shortcomings of the bag model. Instead, I will try to motivate the development of a possible "derivation" of the model from the fundamental theory of strong interactions, that is, from Quantum Chromodynamics.

I shall first present some arguments which suggest a particular variational ansatz for the light quark wave functions of QCD. Next I will discuss what the related gluon wave functions should be, and what the vacuum wave function is when expressed in terms of the amplitudes of these gluon wave functions. Finally, for the main part of the talk, I will derive a finite version of a low energy theorem[3] which relates the mass of the pseudoscalar, flavor singlet meson (the η') to the dependence of the vacuum energy of quarkless QCD on the "θ" parameter, or so-called vacuum angle. I will then apply this theorem to test the model proposed for the gluon vacuum wave function.

II. Bag Model Wave Functions and the Light Quark Vacuum State

The vacuum fluctuations of quarks with a mass m take place over a region of space with a size $\leq 1/2m$ and if this is smaller than the confinement distance a perturbative treatment of the vacuum is quite adequate. In practice, this means the vacuum of the heavy quarks, the c, b, t etc., is perturbative. However, the light quarks have masses which are such that the fluctuations described by the perturbative vacuum wave function take place over a distance longer than the confinement scale and since such fluctuations separate colors over a long distance one would expect that the true vacuum wave function for light quarks would be very different from the free Dirac sea of filled plane wave states. The expected modification would be made on the wave functions of the quarks with momenta smaller than 1/R, where R is the confinement scale.

Now let us digress for a moment and discuss the bag model wave functions of the light quarks. In the bag model the quarks move freely over the "bag", a finite region of space. On the surface the Dirac wave functions obey the boundary condition $(i\gamma\cdot n+1)q=0$, where n is the outwardly drawn normal. The bag vacuum (the "empty bag") is a state in which all the negative energy states are filled. The bag wave functions define a complete set of functions for the finite region of space which is the interior of the bag. Physical hadrons are composed of populating the positive energy states with quarks (and/or removing quarks from negative energy states) in such a way that the quarks and/or antiquarks form a color singlet configuration. It is a property of the bag model and all other simple quark models that all transitions including those in which all the quarks disappear (such as $\pi\to\mu+\nu$) are well described in terms of the "valence" quarks. There is no allowance made for any "empty" bag or vacuum quark participation in the process. In the bag model, the "empty bag" is not a physical hadron. This suggests that the Dirac sea quarks (at least those with low momenta) throughout space are formed into empty bags. This of course must be a consequence of the confining background provided by the gluons of QCD.

By filling space with "empty bags" I mean the following. Let us expand the quark field in terms of a complete set of Dirac wave functions which on a grid scale of the order of the confinement scale correspond to localized bag orbitals for the lowest modes but higher modes which rapidly asymptote to plane waves. We define a trial vacuum state by filling the modes where the free particle Dirac Hamiltonian has a negative expectation. In such a trial vacuum there are no long scale or color density correlations, and hence we would expect that in a confining background the Dirac sea would have a lower energy than in the free Dirac sea. This is what is meant by filling space with empty bags.

Now we come to one of the principal shortcomings of the bag model, the sharp boundaries of the bag. Confinement effects should only influence the very long wave length quark wave functions. The quark wave functions which have high momenta should be unchanged. In the case of the bag model wave functions, this is true except near the bag surface. As a consequence, any physical quantity in the bag model which is sensitive to the region of space near the bag surface is to be viewed with caution.*

----

*One such quantity is the change in energy of the Dirac sea produced by replacing the free wave functions by localized or bag model wave functions. This is the Casimir energy.[4] One finds an expression of the form,

$$E = \frac{a}{R} + \frac{b}{t^2} R + cR^2 \tag{F1}$$

where a is a finite constant, and t is a cut-off associated with the surface thickness. The first two terms in (F1) are the result with the conventional methods used to compute the Casimir energy which are restricted to the use of the Hamiltonian defined on the finite domain. The third term is more subtle. Its origin can be understood with a simple one-dimensional example. Let S be a neutral scalar field and consider the Hamiltonian

$$H = \frac{1}{2}\int dx(\nabla S)^2 + \frac{1}{2}\int dx(\pi)^2 + k(\int_{a-\varepsilon}^{a+\varepsilon} dxS)^2 + 1(\int_{b-\varepsilon}^{b+\varepsilon} dxS)^2 \tag{F2}$$

Then compute the expectation of the free Hamiltonian in the ground state of H. This state should approximate the kind of ansatz we are making for a trial wave function in the limit as $k,l\to\infty$. In this limit one obtains,

$$\Delta E = -\frac{\pi}{24}\left(\frac{1}{|b-a|}\right) + \frac{c_0}{\varepsilon}(\log(k\ell)) \tag{F3}$$

The first term is that gotten by the standard type of computation of a Casimir energy. The last term is the result of enforcing the constraints $S(a)=0, S(b)=0$ on the term $\frac{1}{2}\int dx(\pi)^2$ in the energy. A similar result is anticipated in three dimensions.

Therefore, to improve the model one needs a set of wave functions which on a coarse grained scale look locally like bag model wave functions, without the sharp boundaries, and which are complete over all space. We have suggested that the "Wannier" wave functions used in solid state physics would provide such a set.[5] One probably doesn't need to be so fancy and any complete set formed using localized orbitals will probably be a good basis to describe an ansatz for the light quark vacuum state. I will not discuss this in more detail here.

III. The Gluon Wave Functions in QCD[5]

Just as confinement means that we expect a strong modification of the low momentum components of the light quark Dirac sea, we should also expect that it self-consistently would mean that the amplitude of long wave length gluons would be strongly suppressed in the vacuum wave function. For otherwise as for the light quarks there would be a high probability of fluctuations which would separate colors a long distance. However, gluonic variables are subtle in that local gauge invariance means that a reasonable trial function should to a good approximation be invariant. In the "canonical" formalism ($A^0=0$ or temporal gauge) local space dependent gauge transformations are generated by the operator,

$$G^a(x)=\frac{1}{g_s}\nabla\cdot E^a(x) + f^{abc}\, E^b(x)\cdot A^c(x)$$

Local gauge invariance means $[H,G^a(x)]=0$, and $[G^a,G^b]=f^{abc}G^c(x)\cdot\delta^{(3)}(x-y)$.
If we divide space on the coarse grained scale into localized orbitals which occupy regions R, then the quasi-local generators

$$G^a_R = \int_R d^3x G^a(x)=\frac{1}{g_s}\int dSn\cdot E^a(x) + f^{abc}\int_R d^3x E^b(x)\cdot A^c(x)$$

can be made to vanish on a trial vacuum wave function by making the first term zero by a proper choice of the gluon wave functions in terms of which we expand the field and the second term will vanish when acting on a vacuum wave function in which only total color singlet modes of the field are occupied in each cell R. If we expand the field in terms of suitably defined Wannier functions both of these features can be achieved. Although the simple sharp boundary bag model wave functions are not adequate to compute the trial vacuum energy of all of the "unoccupied" modes of the trial wave function, they are still useful in estimating the energy contribution from low lying occupied modes since this energy is insensitive to the sharp bag walls. In this way we found that it is favorable to populate the ground state with a color singlet gluon pair in each cell into which space has been subdivided. That is, we found that we would expect that the vacuum state of QCD would consist of a condensate of non-overlapping gluon pairs. Since the vacuum is filled with such pairs one would expect that the first excited state (the 0^{++} glueball) would be a particle where in place of one pair there is a hole in the gluon sea. It is this proposal for the vacuum wave function which we should like to test in the next section.

IV. The UA(1) Sector of QCD[6]

Some years ago[3] an analysis of the so-called "UA(1)-problem" was made in the limit of large N (=number of colors). It was shown that in this limit the mass of the ninth pseudoscalar meson should vanish and a formal expression for the squared mass (of order 1/N) was given. Here a version of this argument will be given which produces an expression for the squared mass which can be used for phenomenological purposes. Such an application will then be made to test the model for the QCD vacuum wave function which has been described earlier. Only the theory in the limit of massless quarks will be considered.

First the argument given earlier[3] will be briefly repeated with additions made to it as necessary along the way to produce the final revised form of the expression for the meson mass. In QCD with at least one massless quark there can be no "θ" dependence of the vacuum energy.[7] Here "θ" is the parameter in the formal term in the QCD Lagrangian, $\theta g^2/32 \cdot \sum_a F^a_{\mu\nu} \tilde{F}^a_{\mu\nu} = \theta q(x)$, where $\tilde{F}_{\mu\nu} = \frac{1}{2}\epsilon_{\mu\nu\alpha\beta}F^{\alpha\beta}$. This is because the formal term can be removed from the Lagrangian by making a global chiral transformation on one or more of the massless quark fields. Therefore,

$$\left.\frac{d^2E}{d\theta^2}\right|_{\theta=0} = -\int d^4x \; i\langle T^*(q(x)q(0))\rangle \tag{1}$$

where E = energy per unit volume of the vacuum state, must vanish. Equation (1) may be expressed in terms of the two point function for the composite operator q(x),

$$U(Q) = \int d^4x e^{-iQ\cdot x} \; i\langle T^*(q(x)q(0))\rangle \tag{2}$$

so

$$\frac{d^2E}{d\theta^2} = -U(0) \tag{3}$$

and hence for massless quarks (1) is equivalent to U(0) = 0. In (2), U(Q) is defined as the vacuum expectation of a T* product of the composite operators q(x), that is we assume here that whatever "contact" terms are required to make (1) a correct expression for the vacuum energy dependence on "θ" have been included in the definition of U(Q). Since $q(x) = \frac{g_s^2}{32\pi^2}\sum_a F_a\tilde{F}_a$ is invariant under renormalization the operator should have its "naive" dimension which is four and the dispersion representation of U(Q) must be subtracted three times, so

$$U(Q) = -(Q^2)^3\int \frac{dm^2}{(m^2)^3}\frac{\sigma(m^2)}{Q^2+m^2} + U(0) + U'(0)Q^2 + \frac{1}{2}U''(0)(Q^2)^2 \tag{4}$$

Now from the operator product expansion for q(x) it follows that as $Q^2 \to \infty$,[7]

$$\int d^4x \; e^{-Q\cdot x} iT^*(q(x)q(0)) \to (Q^2)^2 \left\{ \frac{\alpha_s(Q)^2}{32\pi^4} \log \frac{Q^2}{\Lambda^2} + C \right\} - \frac{\alpha_s(Q)}{16\pi}\cdot\left(\frac{\alpha_s}{\pi}\sum_a F_a^{\mu\nu}F_a^{\mu\nu}\right)_{ren} + \text{higher twist} \tag{5}$$

Here C is the arbitrary subtraction constant associated with the vacuum expectation of the product of operators in (1).

It follows from this and asymptotic freedom ($\alpha_s(Q) \to 0$, as $Q^2 \to \infty$) that there can be no terms of the type represented by the subtraction constants U(0) or U'(0) present in the asymptotic form of U(Q), so U(0) and U'(0) must be cancelled by terms which come from the limiting form of the dispersive part of $U(Q)$, and thus it must be possible to express the subtraction constants in terms of the absorptive part $\sigma(m^2)$. The subtraction $U''(0)$ is arbitrary and represents the primitive divergence in the Green's function $U(Q)$, that is, there is a simple relation between C and $U''(0)$.

To isolate the polynomial part of the asymptotic version of the first term of (4) it is convenient to use the representation

$$\frac{(Q^2)^3}{(m^2)^2(Q^2+m^2)} = -\frac{1}{2\pi i}\int_{-i\infty-2-\varepsilon}^{i\infty-2-\varepsilon} d\nu \, \frac{\pi}{\sin\pi\nu}\left(\frac{m^2}{Q^2}\right)^{\nu} \tag{6}$$

valid for $0 < \varepsilon < 1$. If this representation is inserted into (4) and the polynomial terms are picked out as $Q^2 \to \infty$, we find that

$$U(0) = \mathrm{Re}(S(0)) \tag{7}$$

where the analytic function,

$$S(\nu) = \int \frac{dm^2}{m^2} (m^2)^{\nu}\, \sigma(m^2) \tag{8}$$

is defined for $\mathrm{Re}(\nu) < -2$. An expression for $U'(0)$ may also be obtained but shall not be considered here. The expression for $U(0)$ originally obtained[3] was given formally as

$$U(0) = \int \frac{dm^2}{m^2}\, \sigma(m^2) + C' \tag{9}$$

where no explicit form for the constant C' was given in terms of $\sigma(m^2)$. Equation (7) may be regarded as a "renormalized" version of (9). Implicit in the use of the operator product expansion given above is the assumption that it has the proper form to represent the formal transformation which transfers the vacuum angle "θ" into the effective term in the Lagrangian, $\theta q(x)$.*

In evaluating the operator product the definition of the T product which gives (5) is that which makes $i\langle T^*(F_a^{\mu\nu}(x)F_b^{\alpha\beta}(0))\rangle$ in zero'th order take the form in momentum space $(g^{\mu\alpha}q^{\nu}q^{\beta}+\cdots)\frac{1}{q^2}$. There is no contact term of the form $g^{\mu\alpha}g^{\nu\beta}-g^{\mu\beta}g^{\nu\alpha}$. This is because when all the indices have space values, the T* product should be equivalent to the T product if we are to correctly apply covariant perturbation theory based upon the interaction Lagrangian $\theta F\tilde{F}$. This violates naive duality since this form of the T* product gives a contact term in $\langle T^*(\tilde{F}\tilde{F})\rangle$.

The use of an analytic continuation in (8) is closely related to the use of a dimensional regulation to achieve the perturbative terms in the operator product expansion. However the analytic continuation applies in a more general context.

The low energy theorem $U(0)=0$, in the large N limit becomes an expression for the mass of the lowest quark flavor singlet pseudoscalar meson. This is because as $N\to\infty$, the pure gluon contributions to $U(Q)$ are of one whereas the quark state contributions are of order $1/N$ for any non-vanishing value of Q. However, the quark contributions must cancel the gluon contributions to make $U(0)$ vanish. This can happen only if there is a quark meson state with a squared mass of order $1/N$ to enhance the quark contribution to $U(0)$. Thus with $U(0)$ of the form,

$$0 = \mathrm{Re}\,[S_{gluon}(0)] + \frac{g_Q^2}{m_{\eta'}^2} \tag{10}$$

where the lowest quark resonance contribution has been made explicit with

$$g_Q = \langle 0|q(x)|\eta'\rangle.$$

One can express g_Q^2 in terms of the axial decay constant by using the equation for the anomaly of the UA(1) axial current,

$$\langle 0|U^\mu(x)|\eta'\rangle \equiv i\, q^\mu \sqrt{N_f}\,\sqrt{2}\,F_{\eta'}\, e^{iq\cdot x}$$

($N_f = 3$ = #of flavors). Since

$$\partial_\mu U^\mu(x) = 2N_f q(x)$$

one finds that

$$g_Q = \frac{F_{\eta'} m_{\eta'}^2}{\sqrt{2N_f}}$$

This representation shows explicitly that as $N\to\infty$, $m_{\eta'}^2 \sim \frac{1}{N}$, since the first term of (10) is of order 1 and $F_{\eta'}^2$ is of order N. Above we have normalized F so that in the large N limit $F_{\eta'} = F_\pi$ (= 93 MeV). Therefore, we have the sum rule,

$$\frac{m_{\eta'}^2 F_{\eta'}^2}{6} = -\mathrm{Re}[S_{gluon}(0)] = \left(\frac{d^2E}{d\theta^2}\right)_{\text{No Quarks}} \tag{11}$$

The sum rule (11) may be equivalent to one of those derived by the non-perturbative QCD group[8] but I believe that the above derivation is sufficiently different to be interesting. In any case we shall exploit it in a distinct way.

There are a number of ways of looking at (11). We shall limit ourselves to one remark. In perturbation theory $U(Q)$ corresponds to the expectation of (5) and hence because $\langle F^2\rangle=0$, $U(Q) \sim (Q^2)^2(\ldots)$ and so $U(0)=0$ consistent with the absence of any θ dependent results in perturbation theory. Thus, we should also get zero for $Re(S_{pert.}(0))$. In perturbation theory the absorptive part is $\sim\sigma_0 m^4$ and the spectrum of states starts from zero. Hence the integral which defines $S(\nu)_{pert.}$ would diverge at the lower limit. We recognize this to be closely connected to the primitive divergence related to the subtraction constant C. In perturbation theory this problem is resolved by using an infra-red cut-off λ on the lower limit in (8), and then defining S(0) to be the result one gets as the cut off is removed. This gives in lowest order $S(\nu) = -\frac{\sigma_0}{\nu+2}(\lambda^2)^{\nu+2}$, so $S(0) = -\frac{1}{2}\sigma_0\lambda^4 = 0$ as $\lambda\to 0$, consistent with U(0) as calculated directly. In non-perturbative QCD there is a mass gap and no such device is needed. Viewed in this way it is the mass gap which makes U(0) non-vanishing.

It is perhaps not necessary to caution that as it stands the value of S(0) is sensitive to the place where we join $\sigma(m^2)$ to the perturbative form given by asymptotic freedom and the operator product formula based on (5). However to get at least a rough evaluation we shall put

$$S(\nu) = \int \frac{dm^2}{m^2}(m^2)^\nu \sigma(m^2) = \int_{m^{*2}}^{\infty} \frac{dm^2}{m^2}(m^2)^\nu \sigma_{pert.}(m^2) + \int^{m^{*2}} \frac{dm^2}{m^2}(m^2)^\nu \sigma(m^2) \tag{12}$$

where $\sigma_{pert}(m^2)$ for m^2 sufficiently high is given by evaluating $U(Q^2)$ for Q^2 large. We then obtain

$$U(0) = Re(S(0)) = \int^{(m^*)^2} \frac{dm^2}{m^2}\sigma(m^2) - \frac{\alpha_s^2(m^*)}{64\pi^2}(m^*)^4\,\Delta(m^*/\Lambda)^4 - \alpha_s(m^*)\langle\frac{\alpha}{\pi}F^2\rangle_{ren.}\left(\frac{1}{16\pi}\right) \tag{13}$$

where $\Delta(\frac{m^*}{\Lambda})$ is within a few percent of unity over the interesting range of m^*/Λ. Hence the sum rule takes the form

$$\frac{m_\eta^2 F_\eta^2}{6} + \int^{m^{*2}} \frac{dm^2}{m^2}\sigma(m^2) = \frac{\alpha_s^2(m^*)}{64\pi^2}(m^*)^4 + \frac{\alpha_s(m^*)}{16\pi}(\langle\frac{\alpha}{\pi}F^2\rangle)_{ren.} \tag{14}$$

For simplicity, to apply (11) to our vacuum model let us fix m^* by the condition that the first terms on both sides are equal and then evaluate the second terms using our vacuum model for the left side and the quoted phenomenological value[7] of $\langle\frac{\alpha_s}{\pi}F^2\rangle_{ren} = .04\ \text{GeV}^4$ for the right side [we shall use $\Lambda = 200$ MeV.] This gives for m^* about 3.2 GeV using $m_{\eta'} = .95$ GeV and $F_\eta = F_\pi = .095$ GeV. This is not an unreasonable value for m^*. Therefore we get

$$\int^{m*} \frac{dm^2}{m^2} \sigma(m^2) = (.12)^4 \text{ GeV}^4 .$$

Now let us compute the lefthand side using the vacuum model which we have advocated earlier. There is a pseudo-scalar glueball which we take to be the "iota", $m_p \cong 1.5$ GeV. If we assume that the lowest resonance dominates we get

$$\left(\frac{1}{m_p^2} g_p^2\right)_{\text{"exp"}} = (.12)^4 \text{ GeV}^4 \tag{15}$$

where $g_p = \langle 0|q(x)|p\rangle$. We have evaluated g_p using the naive bag model for the pseudoscalar glueball and two possible "bag models" for $|0\rangle$. In the first case there are no valence gluons in the vacuum, and in the second case, the vacuum is filled with scalar gluon pairs. We obtain in these two cases the following expression for g_p^2,

$$g_p^2 = \left[\alpha_s^2 \frac{m_p}{R^5}\right] \begin{bmatrix} 5.75 \\ .158 \end{bmatrix} \begin{bmatrix} \text{Case 1} \\ \text{Case 2} \end{bmatrix} \tag{16}$$

where R is the bag radius. We may take[9] $R \simeq 3.5 \text{ GeV}^{-1}$ and $m_p \simeq 1.5$ GeV if we believe that the "ι" is a pseudoscalar glueball. We have taken $\alpha_s \sim 1$. In this case we find

$$\left(\frac{1}{m_p^2} g_p^2\right)_{\text{"theor"}} = \begin{bmatrix} (.16)^4 \text{ GeV}^4 \\ (.49)^4 \text{ GeV}^4 \end{bmatrix} \begin{bmatrix} \text{Case 1} \\ \text{Case 2} \end{bmatrix} \tag{17}$$

We see that the "filled" (case 2) vacuum is a far better model than the "empty" one (case 1). However R is probably too small as is maybe m_p.

V. Conclusions

I have used a form of the low energy theorem in the pseudoscalar sector of QCD as support for a model for the QCD vacuum based on a condensate of scalar glueballs.

VI. Acknowledgements

I would like to thank Hans Hansson and Bill Bardeen for valuable help. This work was supported in part by funds provided by the U.S. DEPARTMENT OF ENERGY (DOE) under contract DE-AC02-76ER03069.

References

1.) A. Chodos, R. L. Jaffe, K. Johnson, C. B. Thorn, Phys. Rev. D10, 2599 (1974); T. DeGrand, R. L. Jaffe, K. Johnson, J. Kiskis, Phys. Rev. D12, 2060 (1975).

2.) For a recent review see J. Donoghue, C. DeTar; Annual Review of Nuclear Science and Particle Science, Vol. 33 (1983).

3.) E. Witten, Nucl. Phys. B160, 57 (1979).

4.) Some references to related calculations are H. B. G. Casimir, Proc. K. Ned. Akad. Wet. B51, 793 (1948); K. A. Milton, Phys. Rev. D22, 1441 (1980); C. Peterson, T. H. Hansson, K. Johnson, Phys. Rev. D26, 415 (1982); K. A. Milton, Phys. Rev. D27, 439 (1983).

5.) T. H. Hansson, K. Johnson, C. Peterson, Phys. Rev. D26, 2069 (1982).

6.) There are many references to the UA(1) problem. For closely related work see M. A. Shifman, A. I. Vainstein, V. I. Zakharov, Nucl. Phys. B147, 385, 448 (1979); S. Narison, Phys. Letters, 125B, 501 (1983), and other references contained therein.

7.) R. Jackiw, C. Rebbi, Phys. Rev. Lett. 37, 172 (1976); C. G. Callan, R. F. Dashen, D. J. Gross, Phys. Rev. Lett. 63B, 334 (1976); For a pedagogical discussion of "θ" see "Quarks Leptons & Gauge Fields", K. Huang, WORLD SCIENTIFIC, Singapore (1982).

8.) V. A. Novikov, M. A. Shifman, A. I. Vainstein, V. I. Zakharov, Nucl. Phys. B191, 301 (1981).

9.) C. Carlson, T. H. Hansson, C. Peterson, Phys. Rev. D27, 1556 (1983); T. H. Hansson, private communication.

SELF ENERGY OF MASSLESS QUARKS IN THE MIT BAG

J. D. Breit
The Institute for Advanced Study
Princeton, New Jersey 08540

1. Introduction

In this talk I would like to tell you about a calculation of the self energy of a massless quark in the MIT bag [1]. The motivation for doing this calculation goes beyond simply finding small corrections to the hadron masses predicted by the bag model. The bag model gives reasonable results for the masses only when an ad hoc "zero-point" energy is added, that is,

$$M = N\omega + N\alpha\vec{\sigma}_1\cdot\vec{\sigma}_2\ I_{mag} + \frac{4}{3}\,p\pi R^3 - Z_0/R$$

where N is the number of quarks in the bag, ω the kinetic energy per quark--2.04/R for massless quarks, I_{mag} the energy of interaction between quarks due to exchange of a transverse gluon--0.12/R for massless quarks, α the fine structure constant--usually taken to be about 2.2, R the bag radius, p the bag pressure, and $-Z_0/R$ the zero-point energy [2]. This negative zero-point energy was introduced in analogy with the attraction of parallel conducting plates; unfortunately all attempts to find a similar negative energy in the bag have failed. There is a good physical reason for this failure: this zero-point energy causes an empty bag to have energy $\frac{4}{3}\pi pR^3 - Z_0/R$ and hence makes the vacuum unstable.

A way of fitting the mass spectrum without introducing a $-Z_0/R$ term was proposed by Friedberg and Lee [3]. They noticed that the Coulomb interaction between quarks is large and negative, $-0.85N\alpha/R$ for massless quarks, and can therefore fill the same function as the zero-point energy. Because it is proportional to N, it disappears if the bag is empty and so causes no problems with vacuum stability. For completeness, we should include all the $O(\alpha)$ radiative corrections, that is, we should calculate the self-energy. If the self energy is less than $0.85\alpha/R$, we will still have a negative N/R term; if it is greater than $0.85\alpha/R$ we do not and must look elsewhere for an explanation of the origins of the zero-point energy. So even an approximate calculation of the self energy is quite important. If it is small we have a plausible explanation of what has been a serious defect in the bag model.

The calculation of the self energy is divided into two parts. First we use the fixed-cavity approximation, in which the bag does not change as the quarks are excited. The approximation has a serious flaw, which we correct in the next section.

2. Fixed cavity self energy

To calculate the self energy we need propagators for quarks and gluons satisfying MIT boundary conditions at the walls of a fixed spherical cavity. The derivation of the propagators in closed form as a sum over partial waves is straightforward. Since the resulting expressions are rather lengthy, I will not show them here. They can be found in Reference 1 and many other places. Some prefer to split off the free part of the propagators in order to make renormalization more convenient [4,5], but renormalization is straightforward in either case so this is purely a matter of taste. Propagators written as mode sums rather than in closed form can be found in Reference 6.

Let me say a few words about renormalization. Since we are looking at massless quarks, the self energy is finite, and there is no need to introduce any counterterms. The easiest way to see this is to regard the bag as a fixed external scalar field ϕ interacting with the quarks through a term $g\bar{\psi}\psi\phi$ in the Lagrangian and let $g \to \infty$. The possibly divergent graphs are those with a free gluon and either a free quark or a quark that interacts only once with the external field. The first graph just gives the usual free particle mass renormalization, which vanishes for massless quarks. Note that no finite renormalization should be added because we want the limit $R \to \infty$ to reproduce the usual free particle results. The divergent part of the second graph vanishes in the limit $g \to \infty$ because it is proportional to $\bar{\psi}(R)\psi(R)$, which is zero by the MIT boundary condition.

We calculate the self energy by first integrating over angles, leaving a sum over the total angular momentum J, making a Wick rotation in the loop energy ν, integrating numerically over r,r', and ν, and then summing over J. The results for the first five values of J are given in the table in units of α/R.

Table 1. Contributions to the self energy

J	$E_{Coulomb}$	E_{TE}	E_{TB}	E_{Total}
1/2	0.70	-0.50	-0.10	0.10
3/2	0.71	-0.16	-0.09	0.46
5/2	0.35	-0.10	-0.07	0.18
7/2	0.22	-0.08	-0.05	0.09
9/2	0.16	-0.06	-0.05	0.05

I would estimate these results to be good within a few percent. Because we have made a Wick rotation the integrands do not oscillate; so there should be no large source of error in the numerical integration routines.

Finally we are left with a sum over J. To estimate the contribution of J > 9/2 to the series we fit the J = 7/2 and 9/2 results to a function of the form $A/J^2 + B/J^3$. Adding this function to the values in the table gives a self energy of approximately $1.1\alpha/R$, where I would estimate the error due to this procedure to be about 10%. Chin, Kerman, and Yang obtained a self energy about 15% larger than mine [6]. Although the calculated terms to much higher J, they used a mode sum instead of an integral over the loop energy, and that sum converges quite slowly.

You will recall that we need a self energy of less than $0.85\alpha/R$ in order to mimic the effects of the zero point. Although more accurate calculations can and should be done [4,5], they will not alter the fact that the self energy is much too large. So we must look for another source of error if our program is to succeed.

3. Beyond the fixed-cavity approximation

Let us first consider why one might expect the fixed-cavity approximation to be reasonable. Although we know that exciting a quark and gluon increases the pressure in the bag and forces it to expand, for the lowest excited states this expansion is small. Since the self energy is finite, highly excited states, for which the departure from a fixed cavity is large, do not contribute very much. There is, however, one important exception to this picture. If quarks are excited into a p-wave, they have considerable overlap with the translational modes of the quark-bag system, modes that should be excluded from my calculation of the self energy. The contribution of these modes can be quite large; so as a first correction to the fixed-cavity results, we should replace the p-wave quark states with states that are orthogonal to the translational modes.

To find these states, we use the results of Ref. 7, in which the small oscillations of the coupled quark-bag system were quantized. I will illustrate the method for scalar quarks; generalization to spinors is straightforward but tedious. We start with the Lagrangian

$$L = -\partial_u\phi^*\partial_u\phi - \frac{1}{2}(\partial_u\sigma)^2 - g^2\kappa^2\phi^*\phi\sigma^2 - g^2V(\sigma g) .$$

For a carefully chosen potential V and $\kappa \to \infty$, this Lagrangian has baglike solutions,

$$\phi = \frac{1}{\sqrt{2}}\frac{B(x-X)}{g}e^{i\omega t} ,$$

$$\sigma = \frac{A(x-X)}{g} ,$$

where X is the center of mass,

$$B(x-X) = j_0(\omega r),\ \omega = \pi/R,$$

$$A(x-X) = \Theta(r-R) .$$

The charge

$$Q = \frac{i}{\sqrt{2}} \int (\dot{\phi}^*\phi - \dot{\phi}\phi^*)$$

is proportional to $1/g^2$; so expanding in small g is equivalent to demanding that the charge be large. We make the expansion [8]

$$\phi = \frac{1}{\sqrt{2}} \quad \frac{B(x-X)}{g} + q_n(t) \; \beta_n(x-X) \quad e^{-i\xi(t)}$$

$$\sigma = \frac{A(x-X)}{g} + q_n(t) \; \alpha_n(x-X) \; ,$$

where X and ξ are collective coordinates, q_n real quantum coordinates, α_n real c number wave functions, and β_n complex c number wave functions. To exclude translations in X or ξ, we demand

$$\int [\tfrac{1}{2}(\beta_n+\beta_n^*) \; \nabla B + \alpha_n \nabla A] = 0 \; ,$$

$$\int (\beta_n-\beta_n^*) \; B = 0 \; ,$$

and normalize by

$$\int [\tfrac{1}{2}(\beta_n\beta_m^* + \beta_m\beta_n^*) + \alpha_n\alpha_m] = \delta_{nm} \; ,$$

Letting $p_n = \delta L/\delta q_n$, we find the quantum Hamiltonian

$$H = \tfrac{1}{2}[p_n + \frac{i\omega}{2} \int (\beta_n^*\beta_m-\beta_n\beta_m^*) \; q_m]^2$$

$$+ \frac{1}{2} \; q_n q_m \int \beta_n^*(-\nabla^2-\omega^2+\kappa^2 A^2) \; \beta_m$$

$$+ \; q_n q_m \int \alpha_n \kappa^2 AB(\beta_m+\beta_m^*)$$

$$+ \frac{1}{2} \; q_n q_m \int \alpha_n (-\nabla^2+\kappa^2 B^2 + \frac{\partial^2 V}{\partial\sigma^2}) \; \alpha_m$$

$$+ \frac{1}{2} \; q_n q_m \frac{\omega^2}{\int B^2} \left[\int (\beta_n+\beta_n^*) \; B\right]\left[\int (\beta_m+\beta_m^*) \; B\right]$$

$$- \frac{1}{2} \; q_n q_m \frac{\omega^2}{\int [(\nabla B)^2+(\nabla A)^2]} \left[\int (\beta_n-\beta_n^*) \; \nabla B\right]\left[\int (\beta_m-\beta_m^*) \; \nabla B\right]$$

$$+ \; O(g)$$

For g small we drop the O(g) and smaller terms and diagonalize the quadratic Hamiltonian to find the eigenstates and eigenergies.

For fermions the expansion is in the inverse of the number of quarks in the bag, and is therefore similar to the more familiar large-N expansion. The eigenvalues of H can be found exactly; finding the eigenfunctions would require truncating H and generating the functions numerically. Instead we drop a part of H that changes the eigenvalues only slightly, and determine the eigenfunctions of the remaining part of the Hamiltonian exactly. We use these new p-wave eigenfunctions to recalculate the part of the self energy due to the instantaneous Coulomb interaction when the quark is excited into a p-wave. The change in the self energy is quite large; the new result is $\sim 0.65\alpha/R$.

Now the sum of the Coulomb interaction between quarks and the self energy is negative, about $-0.2N\alpha/R$. For $\alpha \sim 2.2$, $N=3$, this is $\sim -1.3/R$, already quite close to the value usually used for the zero point, $-1.8/R$.

4. Conclusions

Although these results are approximate, we can draw two conclusions: First, radiative corrections can provide the bulk of the ad hoc "zero-point" energy included in MIT bag model phenomenology, Second, the fixed-cavity approximation should not be trusted to better than about 50%.

Because it is so important to understand the origins of the $-Z_0/R$ term, these calculations should be carried to higher accuracy. The fixed-cavity calculation is currently being improved upon by calculating higher angular momentum terms [4,5]. More important, however, would be finding a better method, perhaps variational in nature, for removing the translational modes.

Acknowledgment

This work is supported by the Department of Energy under grant No. DE-AC02-76ER02220.

References

[1] J. D. Breit, Columbia University Preprint CU-TP-229 (1982).

[2] T. de Grand, R. L. Jaffe, K. Johnson, and J. Kiskis, Phys. Rev. D 12, 2060 (1975).

[3] R. Friedberg and T. D. Lee, Phys. Rev. D 16, 1096 (1977).

[4] J. Baacke, Y. Igaraski, and G. Kasperidus, Dortmund University preprint DOTH 82/13 (1982).

[5] T. H. Hansson and R. L. Jaffe, MIT preprint CTP 1026 (1982).

[6] S. A. Chin, A. K. Kerman, and X. H. Yang, MIT preprint CTP 919 (1981).

[7] J. D. Breit, Nucl. Phys. B 202, 147 (1982).

[8] N. H. Christ and T. D. Lee, Phys. Rev. D 12, 1606 (1975).

QUARK MODEL CALCULATIONS OF NUCLEON STRUCTURE FUNCTIONS

C. M. Shakin

Department of Physics and Institute for Nuclear Theory
Brooklyn College of the City University of New York
Brooklyn, New York 11210

We present calculations of those structure functions of the nucleon which are measured in deep inelastic electron scattering. A quark model which preserves translational invariance is used. The model exhibits scaling and the structure functions satisfy the Callan-Gross relation in the scaling region. It is possible to fit the experimental values of $F_2^p(x)-F_2^n(x)$ using wave functions that correspond to a relatively small region of confinement. The ratio of $F_2^n(x)/F_2^p(x)$ is also calculated. One can explain the deviation of the value of the latter quantity from the value 2/3 obtained in the simplest quark model by allowing the neutron confinement radius to be about 10 percent larger than the corresponding proton radius.

Almost all theoretical discussion of nucleon structure functions makes use of the parton model.[1] This model has been quite successful in many respects and has provided information concerning the distribution of valence and "sea" quarks in the nucleon infinite-momentum frame. There have been some attempts to relate the structure functions directly to the quark wave functions of the MIT bag model.[2] However, such attempts are limited by the violation of translational invariance in the standard static-cavity version of the bag model.

In this work we attempt to understand the structure functions in terms of quark wave functions. The model used respects translational invariance. We have used a wave function that describes the motion of a quark relative to a diquark structure. (For convenience we use baglike wave functions but we do not make the static-cavity approximation. The wave functions are parametrized by a coordinate of relative motion, as noted above). It is natural to calculate the quantity $F_2^p(x)-F_2^n(x)$ since, in the parton model, this quantity is independent of the "sea" quark distribution.[1] We find that our results are sensitive to two features of the calculation. First, the size of the confinement region plays an important role, and second, the specification of the mass of the quark and the mass of the (spectator) diquark in the final state is quite important.

The details of our calculations may be found in Ref. 3. Our quark wave function is a solution of the free Dirac equation for r less than some confinement radius R. At r=R we apply the bag boundary condition, so that our wave functions are "baglike" except that they are expressed in terms of a relative coordinate. We find that values of $R\simeq 0.6$ to 0.7 fm give a good account of the shape and magnitude of the quantity $F_2^p(x)-F_2^n(x)$. These values of R correspond to the values used in "small bag" models of the nucleon. As noted above, the specification of the mass of the struck quark (m_q) and the mass of the spectator diquark (m_d) is also an important feature of these calculations.[3] We have also investigated the value of $F_2^n(x)/F_2^p(x)$. In the simplest form of the parton model one obtains a constant value of 2/3 for this quantity. The data, however, show a strong variation of this ratio with x. We note that a small increase of the neutron confinement radius with respect to the proton confinement radius (of about ten percent) goes a long way toward explaining the x dependence of $F_2^n(x)/F_2^p(x)$.

Our model for deep inelastic scattering is shown in Figure 1. The photon momentum $q=(\nu,\vec{q})$ is transferred to a quark whose momentum is P-K. Here K is the momentum of a diquark system which is left either in a state with S=0, T=0, or S=1, T=1. The indices p, s_q and

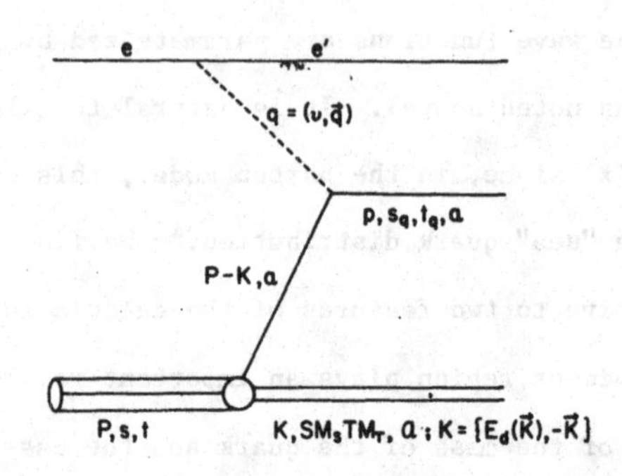

Fig. 1. Deep inelastic electron scattering in the one-photon exchange approximation. The four-momentum of the photon is $q=(\nu,\vec{q})$. The nucleon momentum is P and the spin projection and isospin projections are s and t. The diquark has four-momentum $K=(E_d(\vec{K}),-\vec{K})$, where $E_d(\vec{K})=(\vec{K}^2+m_d{}^2)^{\frac{1}{2}}$. The diquark spin angular momentum is S with projection M_S, etc. The struck quark has momentum P-K before and p=P-K+q after absorption of the photon. The spin and isospin projections of the quark in the final state are s_q and t_q, respectively. Here A and a denote color indices of the diquark and the struck quark.

t_q are the momentum, spin projection and isospin projection of the struck quark. Further, a and A are the color indices of the quark and diquark. (Since the quark belongs to the 3 representation of $SU(3)_c$, the diquark must be in the $\overline{3}$ representation of $SU(3)_c$. Of course, these colored objects are only intermediate states in the evolution of the system in time, as the final state contains only color singlets.) The struck quark is described by a spinor solution of the free Dirac equation, $u(\vec{p},s_q)$, and is assigned a mass, m_q. While the discussion of our wave function in terms of the relative coordinate $\vec{r}$ has some intuitive appeal, the more precise specification of the model is made in momentum space.[3] Equation 3.9 of Ref. 3 provides a parametrization of the product of the nucleon-quark-diquark vertex and quark propagator. The momentum-space wave function is then used to construct the nucleon density matrix. (See Eqs. 2.15 and 2.16 of Ref. 3.)

In the evaluation of the process shown in Figure 1, one must make some specification of the mass, m_q, of the quark and the mass of the diquark, m_d. As we are not dealing with physical channels at this point the choice of these masses is not obvious a priori.

It is our opinion that the natural choice is to assign the struck quark a mass close to zero, the "current" mass as opposed to the "constituent" mass. (The latter number is approximately one-third of the nucleon mass). Our assumption is consistent with the idea that the constituent mass arises from the confinement mechanism and is essentially associated with the low momentum transfer aspects of quantum chromodynamics (QCD). Once the quark is moving with a large momentum one expects that the constituent mass is no longer a relevant parameter. In the MIT bag model the quark has a mass close to zero. In that model the eigenvalue of the Dirac equation plays the role of the constituent mass of the nonrelativistic quark model.

In light of these comments we have performed most of our calculations with $m_q=0$ and $m_d=M_N$. Variation from these values leads to a poor fit in the experimental data.[3]

In Figure 2 we present values of $\nu W_2(Q^2,x)$ and $2xM_NW_1(Q^2,x)$ as a function of Q^2. (Here x=0.25, m_q=5 MeV, m_d=934 MeV, and R=0.6 fm). The independence of Q^2 exhibited for the larger values of Q^2

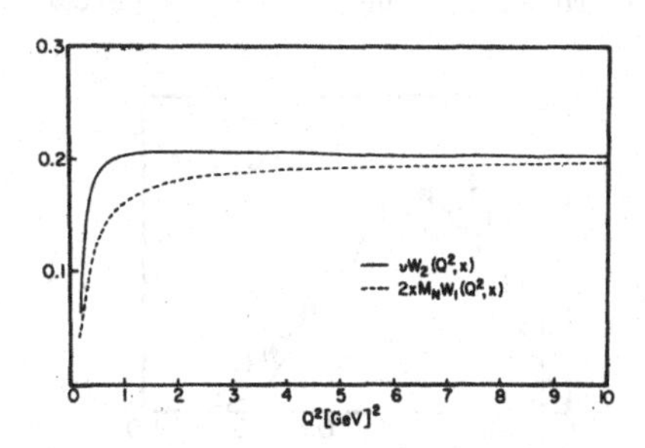

Fig. 2. The solid line represents the values of $\nu W_2(Q^2,x)$ while the dashed line represents $2xM_NW_1(Q^2,x)$. In the scaling limit $\nu W_2(Q^2,x)\to F_2(x)$ and $2xM_NW_1(Q^2,x)\to 2xF_1(x)$. The calculation is made for x=0.25, $R_p=R_n$=0.60 fm, m_q=5 MeV, and m_d=934 MeV.

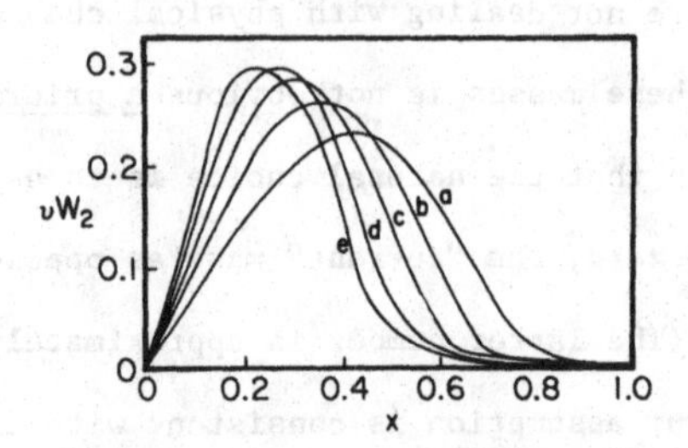

Fig. 3. Values of $\nu W_2(Q^2,x)$ as a function of $x=Q^2/(2M_N\nu)$ calculated for $Q^2=6.25$ $(GeV)^2$. The calculation is made for the wave function parameter $x_0=2.05$ and various values of R: (a) R=0.5 fm, $m_q=6.0$ MeV; (b) R=0.7 fm, $m_q=4.26$ MeV; (c) R=0.9 fm, $m_q=3.31$ MeV; (d) R=1.1 fm, $m_q=2.71$ MeV; and (e) R=1.3 fm, $m_q=2.30$ MeV.

corresponds to "Bjorken scaling." (Note that $F_2(x)$ and $F_1(x)$ are defined such that $\nu W_2(Q^2,x)\rightarrow F_2(x)$ and $M_N W_1(Q^2,x)\rightarrow F_1(x)$ for large Q^2). We see that the Callan-Gross relation, $2xF_1(x)=F_2(x)$, is satisfied for the larger values of Q^2.

In Figure 3 we present values of $\nu W_2(Q^2,x)$ calculated for $Q^2=6.25$ $(GeV)^2$ and wave-function parameter $x_0=2.05$. The variation of $\nu W_2(Q^2,x)$ with R is shown. Here m_q varies from about 2 to 6 MeV and $m_d=M_N-m_q$. (See the caption to Figure 3).

We now turn to a discussion of Figures 4-6. We remark that for the case in which the proton confinement radius parameter, R_p is

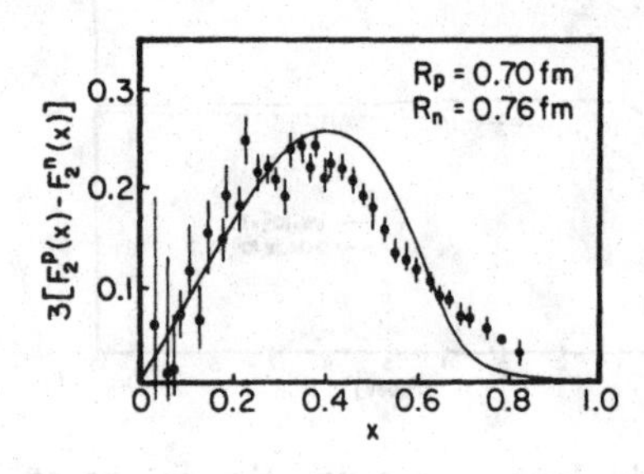

Fig. 4. Values of $3(F_2^p(x)-F_2^n(x))$ are presented as a function of $x=Q^2/(2M_N\nu)$ and compared with experimental data. The calculation is made for $R_p=0.70$ fm, $R_n=0.76$ fm, and $Q^2=6.25$ $(GeV)^2$. Here $m_q\simeq0$.

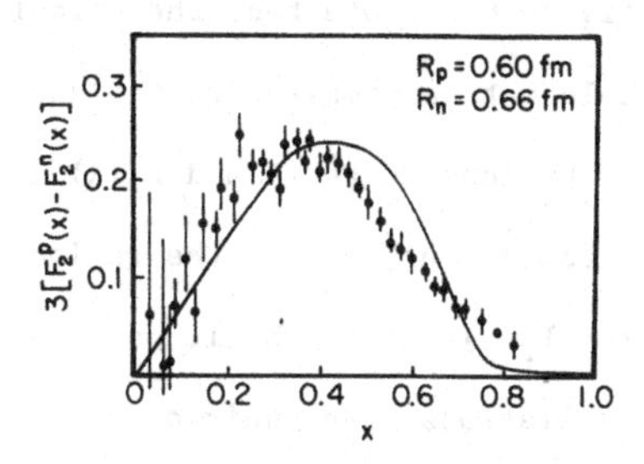

Fig. 5. Values of $3(F_2^p(x)-F_2^n(x))$ are presented as a function of $x=Q^2/(2M_N\nu)$ and compared with experimental data. The calculation is made for R_p=0.60 fm, R_n=0.66 fm, and Q^2=6.25 $(GeV)^2$. Here $m_q \simeq 0$.

equal to the neutron confinement radius parameter, R_n, the value of $F_2^n(x)/F_2^p(x)$ is equal to 2/3. As the data show large variation from this value we have considered the case $R_n > R_p$. Indeed for $R_n - R_p$=0.06 fm one can make a reasonable fit to $F_2^n(x)/F_2^p(x)$. In Figures 4 and 5 we present values of $3(F_2^p(x)-F_2^n(x))$ for R_p=0.70 fm, R_n=0.76 fm and for R_p=0.60 and R_n=0.66 fm, and in Figure 6 we present values for the ratio $F_2^n(x)/F_2^p(x)$ for R_p=0.70 fm and R_n=0.76 fm. These calculations are made for values of $m_q \simeq 0$ and $m_d \simeq M_N$. We see that the fit to the experimental data favors small values of R.

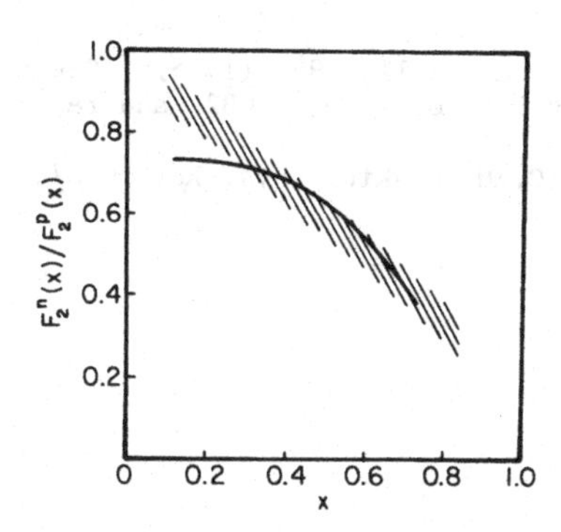

Fig. 6. The ratio $F_2^n(x)/F_2^p(x)$ as a function of $x=Q^2/(2M_N\nu)$. The shaded area represents the data. The calculation is made for R_p=0.70 fm, R_n=0.76 fm, and Q^2=6.25 $(GeV)^2$. Here $m_q \simeq 0$.

With reference to Figure 6 we note that the calculated values of $F_2^n(x)/F_2^p(x)$ probably should not be compared to the data for x<0.2 where the sea quark contributions to $F_2^p(x)$ and $F_2^n(x)$ are either dominant or quite important.[1] Further, in the region beyond x=0.7 our calculated results for $F_2^n(x)/F_2^p(x)$ exhibit oscillations. We have determined that these oscillations have their origin in the sharp radial cutoff of the simple wave functions introduced in Ref. 3. It is clear that these calculations should be repeated with improved wave functions that do not have the coordinate-space discontinuities of our baglike wave functions.

It is interesting to note that our model strongly supports the use of small confinement parameters, R. (The use of such small values of R has been advocated by researchers who have developed the "chiral-bag" model).

References

1. For a review of the parton model, see F. E. Close, An Introduction to Quarks and Partons (Academic, New York, 1979). For a parton-model calculation of structure functions in the proton rest frame, see J. Franklin, Nucl. Phys. B138, 122 (1978); Phys. Rev. D 16, 21 (1977).

2. R. L. Jaffe, Phys. Rev. D 11, 1953 (1975). See also, R. L. Jaffee, MIT Report CTP No. 1029, (1982) and references therein.

3. L. S. Celenza and C. M. Shakin, Phys. Rev. C 27, 1561 (1983).

WHERE HAVE ALL THE PSEUDOSCALAR GLUEBALLS GONE?

William F. Palmer and Stephen S. Pinsky

Abstract

We review the experimental and theoretical status of the $\iota(1440)$ observed in $\psi \to \gamma X$. We comment on previous mass matrix studies and present our own mixing scheme, which can accommodate $\iota(1440)$ either as a radial excitation or as a glueball. To distinguish between the two rates, more data and analysis are needed. In particular, it would be useful to know the spin-parity of the broad object seen at 1710 MeV in $\psi \to \gamma X$.

1. Introduction

$\iota(1440)$ is a $J^P = 0^-$ resonance seen in the $K\bar{K}\pi$ mass distribution in radiative decay of ψ, $\psi \to \gamma K\bar{K}\pi$, with $K\bar{K}$ primarily in the δ state. The mass, width, production, and decay parameters are given by the Mark II and Crystal Ball groups at SPEAR [1-4]:

$$\begin{aligned} &\text{Mass} = 1440 \pm 15 \text{ MeV} \qquad \Gamma = 55 \pm 25 \text{ MeV} \\ &B(\psi \to \gamma\iota)B(\iota \to K\bar{K}\pi) = (4.1 \pm 1.5) \times 10^{-3} \\ &B(\iota \to \delta\pi)B(\delta \to K\bar{K})/B(\iota \to K\bar{K}\pi)' = 0.8 \pm 0.2 \quad . \end{aligned} \tag{1}$$

For comparison, the production of other pseudoscalar mesons in $\psi \to \gamma X$ is

$$\begin{aligned} &B(\psi \to \gamma\eta_c) = (2.0 \pm 3.6) \times 10^{-3} \\ &B(\psi \to \gamma\eta) = (0.86 \pm 0.09) \times 10^{-3} \\ &B(\psi \to \gamma\eta') = (3.6 \pm 0.5) \times 10^{-3} \\ &B[\psi \to \gamma\zeta(1275)] \sim \text{not seen in } K\bar{K}\pi \text{ or } \eta\pi\pi \quad . \end{aligned} \tag{2}$$

The first decay in Eq.(2) is allowed by the OZI rule. The last three are forbidden. [The $\zeta(1275)$, a candidate for the radial excitation of the η, is seen in $\pi^- p \to Xn$][5].

Another state produced prominently in $\psi \to \gamma X$, with spin-parity not inconsistent with $J^P = 0^-$, is the new 1710 MeV resonance seen [6] by the Crystal Ball group in the $\eta\pi\pi$ final state,

New State:

$$\text{Mass} = 1710 \pm 45 \text{ MeV} \qquad \Gamma = 530 \pm 110 \text{ MeV} \tag{3}$$

$$B(\psi \to \gamma\eta\pi\pi) = 5.8 \pm 0.4 \times 10^{-3} \quad .$$

The 1710 is not seen in the $K\bar{K}\pi$ state [1-4]. The 1440 is not seen in $\eta\pi\pi$ by the Crystal Ball [6] group, which has a limit

$$\frac{B(\iota \to \eta\pi\pi)}{B(\iota \to K\bar{K}\pi)} < 0.5 \quad \text{(90\% confidence level)} \quad . \tag{4}$$

η, η', $\zeta(1275)$ and $\iota(1440)$ are thus four established $J^P = 0^-$ isosinglet mesons, and the new state at 1710 MeV is a candidate for a fifth. We display the pseudoscalar masses and properties in Table 1.1.

TABLE 1.1 Summary of the pseudoscalar spectrum

	Mass (MeV)	State	Γ	DECAYS	ASSIGNMENTS
	0				
π					
	200				
K	400				
	600	η (549)	–	–	Ground State ~ η_8
	800				
	1000	η'(958)	–	–	Ground State ~ η_0
π'(1300)	1200	ζ (1275)	70 MeV	$\delta\pi$	Radial excitation of η (η_R) ?
K'(1400)	1400	ι (1440)	55 MeV	$K\bar{K}\pi$	?
	1600				
		NS (1710)	530 MeV	$\eta\pi\pi$	?
	1800				

Conventionally, η and η' are assigned to the ground state nonet of light quark $(\ell\bar{\ell})$ quarkonium. $\zeta(1275)$ has for some time been a candidate for the radial excitation of the η. $\iota(1440)$ has been a potential candidate for (I) a gluonic bound state or glueball [7,8,9] and (II) the radial excitation of the η' [9,10]. If the 1710 MeV resonance has $J^P = 0^-$, it would fit into this scheme as (I) the missing radial excitation of the η' or (II) a glueball. In these proceedings we will assume that there is a fifth pseudoscalar in this mass region, not necessarily the 1710. For the sake of argument we will often use 1710 MeV as the mass of the fifth state, but we do not exclude other masses; indeed our calculations below will allow a range of masses for the fifth pseudoscalar, and one of our models definitely predicts that the 1710 observed by the Crystal Ball group is not the fifth pseudoscalar.

In stating these hypothetical assignments we have assumed that there is no glueball-quarkonium mixing and there is no configuration mixing between ground state and radially excited states of quarkonium. Such mixing is clearly to be expected — or else glue could never decay into quarks; at the very least a gluonic state would mediate the coupling of radial excitations to ground states. The hypothetical assignments thus mean "mostly ground state," "mostly glue," etc. [A third possibility, that the $\iota(1440)$ is a four quark state, will not be considered here.]

How can one begin to choose between a picture closer to assignment (I) or a picture closer to assignment (II)? Let us distinguish three modes of attack:

(A) Production and decay characteristics as measurements of glue versus quarkonium content: the wave function approach.

(B) Spectrum studies correlating masses and mass splittings among quarkonium and glueballs: the mass matrix approach.

(C) A complete dynamical theory in which the mass splittings of (B) are correlated with the wavefunctions of (C), explaining masses and branching fractions in one stroke.

In this report we are primarily concerned with approach (C). However, we would like to make a few observations about approaches (A) and (B).

Approach (A). An argument that $\iota(1440)$ is not the radial partner to $\zeta(1275)$ was first advanced by Chanowitz [10]: since ψ is not observed to decay into $\gamma\zeta$, it must be mostly SU(3) octet in character. If the $\iota(1440)$ is the radial partner to $\zeta(1275)$ it must therefore be mostly SU(3) singlet in character. But then ι should be seen in $\pi^- p \to \iota n$. Since Stanton [5] did not observe this final state in $\pi^- p \to (\delta\pi)n$, there is a contradiction, whence $\iota(1440)$ is not the radial partner to $\zeta(1275)$. Let us however examine this argument more quantitatively. The ratio of ζ production to ι production in $\psi \to \gamma X$, with subsequent decay in $K\bar{K}\pi$, is given by

$$R_1 = \frac{B[\psi \to \gamma\zeta(1275)]B(\zeta \to K\bar{K}\pi)}{B(\psi \to \gamma\iota)B(\iota \to K\bar{K}\pi)} = 0.86 \tan^2\theta_R \tan^2(\theta_R + 54.7°) \quad . (5)$$

Here the factor of 0.86 represents phase space; $\tan^2\theta_R$ is the ratio of the singlet content of ζ to the octet of ι; and $\tan^2(\theta_R + 54.7°)$ is the ratio of the light quark content of $\zeta(1275)$ to the light quark content of $\iota(1440)$. The light quark content is taken because both ζ and ι are known to decay via the light quark $\delta\pi$ state.

Similarly, the ratio of ι to ζ production in $\pi^- p$, with subsequent decay into $\eta\pi\pi$ is given by

$$R_2 = \frac{\sigma(\pi^- p \to \iota n)B(\iota \to \eta\pi\pi)}{\sigma(\pi^- p \to \zeta n)B(\zeta \to \eta\pi\pi)} = 1.33 \tan^4(\theta_R + 54.7°) \quad . \quad (6)$$

With a radial nonet mixing angle $\theta_R = -18°$ we find $R_1 = 0.05$ and $R_2 = 0.4$. Both R_1 and R_2 are within the experimental limits. Thus this simple wave function can not rule out $\iota(1440)$ as a radial excitation.

On the other hand, we have shown in previous publications that the production and decay of $\iota(1440)$ is well described by a pole model in which $\iota(1440)$ is a glueball that mediates OZI violating processes (like $\psi \to \gamma\eta'$.) For details refer to reference 9.

Approach (B). Two groups, Donoghue and Gomm (DG) [11] and Fishbane, Karl and Meshkov (FKM) [12] have looked at the pseudoscalar energy levels.

DG argue that $\iota(1440)$ is not a radial excitation. They base their conclusion on a two channel mass matrix in which $\zeta(1275)$ and $\iota(1440)$ are at first assumed to be descended from ideally mixed radial excitations of η and η'. In a light quark, strange quark basis, their mass matrix is

$$\begin{pmatrix} m^2_{\eta_r} + 2\epsilon & \sqrt{2}\,\epsilon \\ \sqrt{2}\,\epsilon & m^2_{\eta'_r} + \epsilon \end{pmatrix} \begin{matrix} \ell\bar{\ell} \\ s\bar{s} \end{matrix} \tag{7}$$

The bare masses satisfy the ideal mixing mass formula

$$\begin{aligned} m^2_{\eta_r} &= \mu^2_{\pi'} = (1300)^2 \\ m^2_{\eta'_r} &= 2\mu^2_{k'} - \mu^2_{\pi} = 2(1440)^2 - (1300)^2 = (1500)^2 \end{aligned} \tag{8}$$

Since the isotriplet candidate for the radial nonet, $\pi'(1300)$ is so close to $\zeta(1275)$, deviations from ideal mixing (requiring $m_{\pi'} = m_\zeta$) are severely limited, requiring ϵ to be small. This in turn requires the radial excitation of the η' to be at 1530 MeV. They conclude that $\iota(1440)$ can not be the radial excitation of η'.

If, however, the $\iota(1440)$ is not the radial excitation, surely it must mix with the nearby states $\zeta(1275)$ and the nonet partner to ζ, the ζ' (ζ' presumably is near 1530 before mixing begins); but then the masses will be shifted, weakening the DG case against $\iota(1440)$ as largely a radial state.

This leads to the question: how closely can one expect to predict masses in these simple two channel models? DG [11] apply the same two channel model to the ground states, η and η', incorporating the large deviation from ideal mixing by a substantial gluon annihilation term. It turns out, however, that the masses of η and η' are not well represented in this model (quadratic mass matrix: $m_\eta = 495$ MeV, $m_\eta = 970$ MeV). To be fair, chiral symmetry effects result in very large mass differences in the ground state nonet, probably larger than those in the radially excited nonet. However, in view of these numerical results, it seems premature to rule out $\iota(1440)$ as a radial excitation. The conclusion we take from their model is that the $\zeta(1275)$ and $\iota(1440)$ are not simple partners in a pure radial nonet — there must be <u>another</u> state (glueball? radial excitation?) with which ζ and/or ι mix.

FKM [12] use a 3×3 mass matrix, with bare channels of ideally mixed η and η', and a glueball. Free couplings allow SU(3) symmetric mixing of η and η' with the glueball and SU(3) symmetric gluon annihilation diagrams allowing mixing between light ($\ell\bar{\ell}$) and strange ($s\bar{s}$) quarks. When FKM turn off their glueball mixing, they have a 2×2 η, η' mixing matrix identical with that of DG. Their matrix, in ideal mixed basis, is

$$\begin{pmatrix} m_\eta^2 + 2\epsilon & \sqrt{2}\,\epsilon & \sqrt{2}\,\beta \\ \sqrt{2}\,\epsilon & m_{\eta'}^2 + \epsilon & \beta \\ \sqrt{2}\,\beta & \beta & m_G^2 \end{pmatrix} \begin{matrix} \ell\bar{\ell} \\ s\bar{s} \\ \text{Glueball} \end{matrix} \tag{9}$$

where the bare masses satisfy the ideal mixing mass formula

$$\begin{aligned} m_\eta^2 &= \mu_\pi^2 \\ m_{\eta'}^2 &= 2\mu_k^2 - \mu_\pi^2 \quad . \end{aligned} \tag{10}$$

FKM require their two parameter theory to fit η and η' at their observed masses, and predict the third mass. Positivity requirements on the squares of couplings then lead them to a lower limit, 1.7 GeV, for the mass of the third state. They conclude that $\iota(1440)$ is not a glueball. However, they do not accommodate $\iota(1440)$ in their mass spectrum, despite the fact that it must mix with the quarkonium states, or else it could not decay into $K\bar{K}\pi$. It is interesting to note that if FKM decouple their third state, they obtain as good (or bad, depending on viewpoint) a description of the masses of η and η' as DG, but now with the glueball mass completely undetermined. The conclusion that we draw from FKM is that η, η' and ι do not form a dynamically complete system — there must be at least another state with which they mix.

We conclude that neither model is powerful or general enough to decide between glueball or radial status of $\iota(1440)$. Mixing of $\iota(1440)$ with $\zeta(1275)$ would surely be as potentially important as mixing between $\iota(1440)$ and $\eta(558)$. If the new state of 1710 MeV is indeed a pseudoscalar, we must take seriously the mixing between $\zeta(1275)$, $\iota(1440)$ and the new resonance, at least at the 15-20% level.

2. Combined Mass Matrix/Wave Function Approach to Pseudoscalars

We now mix five states, originally diagonal, with quark configurations (channel content) corresponding to

$$
\begin{array}{ll}
\eta_8 & \text{SU(3) octet ground state} \\
\eta_0 & \text{SU(3) singlet ground state} \\
G & \text{Glueball state} \\
\eta_r' & \text{Ideally mixed } (s\bar{s}) \text{ radially excited state} \\
\eta_r & \text{Ideally mixed } \frac{u\bar{u} + d\bar{d}}{\sqrt{2}} \text{ radially excited state}
\end{array}
\tag{11}
$$

Our assumption (quark model) is that all nonets are ideally mixed to begin with except when some principle (in this case the chiral symmetry limit for an octet of Goldstone particles) says otherwise. Thus we start the ground states off in SU(3) singlet and SU(3) octet configuration. We take the bare mass of the octet particle, η_8, to be given by the Gell-Mann Okubo mass formula,

$$m_{\eta_8}^2 = \frac{4\mu_k^2 - \mu_\pi^2}{3} = 0.32 \text{ GeV}^2/c^2 \tag{12}$$

where in our notation μ refers to true mass and m to bare mass (prior to mixing).

The bare mass of the singlet, $m_{\eta_0}^2$, split off from the octet by the axial anomaly contribution, is taken to be a free parameter, m_G^2, the bare mass of the bare glueball.

The bare masses of the ideally mixed radially excited states, $m_{\eta_r}^2$ and $m_{\eta_r'}^2$, are given in terms of the true masses of the isotriplet (π') and strange candidate K' in Eq.(8).

These bare masses are indicated in Table 2.1, along the diagonal and in the 2 x 2 matrix in the lower-right corner. [See following page.] [We write the matrix in the SU(3) singlet and octet basis for the radially excited quark states $(0',8')$ for convenience; the lower right 2 x 2 matrix is diagonal in the ideally mixed bases, obeying Eq.(8).]

The mixing parameters are

f: Glue-singlet ground state mixing

g: Glue-singlet radially excited state mixing

ϵ: Glue-octet ground state mixing

ϵ is a small parameter introduced to account for the small deviation of the physical η mass from the Gell-Mann Okubo formula, Eq.(12).

TABLE 2.1 Mass matrix of the isoscalar pseudoscalars in a SU(3) singlet/octet basis.

8	0	Glue	0′	8′
$\frac{4\mu_K^2 - \mu_\pi^2}{3}$	0	ϵ	0	0
ϵ	$m_{\eta'}^2$	f	0	0
0	f	m_G^2	g	0
0	0	g	$\frac{2\mu_{K'}^2 + \mu_{\pi'}^2}{3}$	$\frac{2\sqrt{2}}{3}\left(\mu_{\pi'}^2 - \mu_{K'}^2\right)$
0	0	0	$\frac{2\sqrt{2}}{3}\left(\mu_{\pi'}^2 - \mu_{K'}^2\right)$	$\frac{4m_{K'}^2 - \mu_{\pi'}^2}{3}$

We thus have a mixing theory with five free parameters (f, g, ϵ, $m_{\eta'}$, m_G). (Other entries in Table 2.1 refer to the physical masses μ_K, μ_π, $\mu_{K'}$ and $\mu_{\pi'}$.) If we require the mass matrix to have the eigenvalues corresponding to the physical η, η', $\iota(1440)$, $\zeta(1275)$, and a high mass pseudoscalar (we take a stab at 1710 GeV) we determine all five parameters and hence can calculate the quark configuration and glue content of the physical states.

This process does not guarantee a solution for all inputs because the outputs must be real, f^2, ϵ^2, $g^2 > 0$. Thus we vary one input parameter in the neighborhood of its physical value: for this purpose we choose the mass of the radially excited strange pseudoscalar, $\mu_R = 1400 \pm 50$ MeV with a width of 250 MeV.

In practice we solve the 4 x 4 system with $\epsilon = 0$ exactly and then handle ϵ perturbatively.

In Figure 2.1 we display f^2 and g^2 as a function of the physical mass of the radial K′. [See following page.] This is equivalent to varying the mass of the radial bare $s\bar{s}$ state. There are several regions where f^2 and g^2 are positive. The physically interesting ones are at (I) $\mu_{K'} \approx 1400$ MeV and (II) $\mu_{K'} \approx 1500$. In these regions the mixing problem has a real solution for f and g, with channel content as shown in Table 2.2. [See following page.]

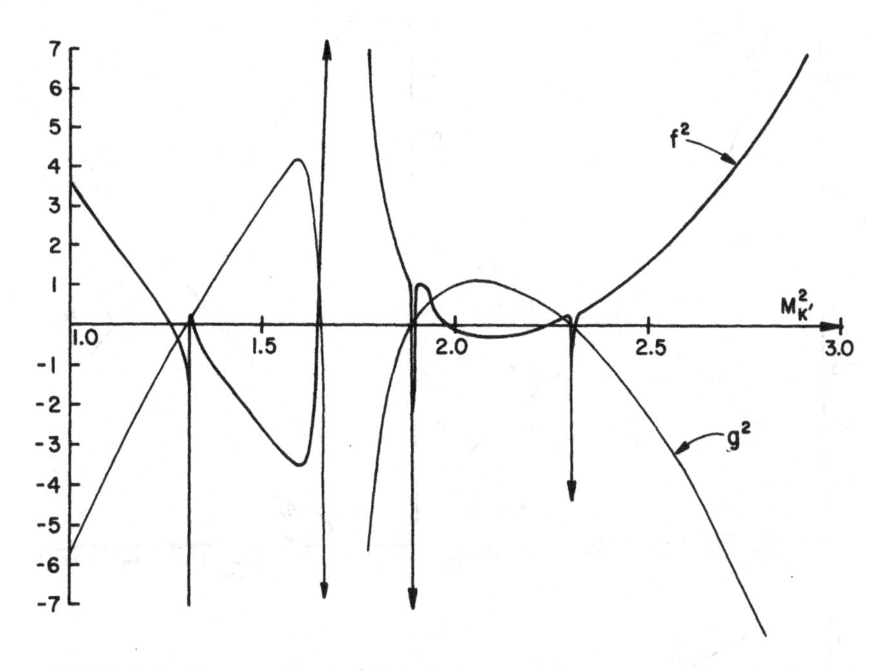

FIGURE 2.1 The coupling (f) of the ground states to the glueball and the coupling (g) of the radially excited states to the glueball as a function of $\mu^2_{K'} = m^2_{K'}$. (The mass of the fifth pseudoscalar is taken to be 1710 MeV.)

TABLE 2.2 Two classes of solution to the 5 x 5 pseudoscalar mass matrix.

State	Solution I $\mu_{K'} \sim 1440$ channel content		Solution II $\mu_{K'} \sim 1500$ channel content	
$\eta'(958)$	η_o		η_o	
$\zeta(1275)$	η_r	$(\ell\bar{\ell})$	η_r	$(\ell\bar{\ell})$
$\iota(1440)$	η'_r	$(s\bar{s})$	Glueball	
NS(1710)	Glueball		η'_r	$(s\bar{s})$

In Figures 2.2(a)-(d) [see following page] we display the square of the wave function for each physical particle in solution I in terms of the channel content of the bare states, Eq.(11). In solution I the $\iota(1440)$ is mostly a radial excitation of η' and a partner to $\zeta(1275)$.

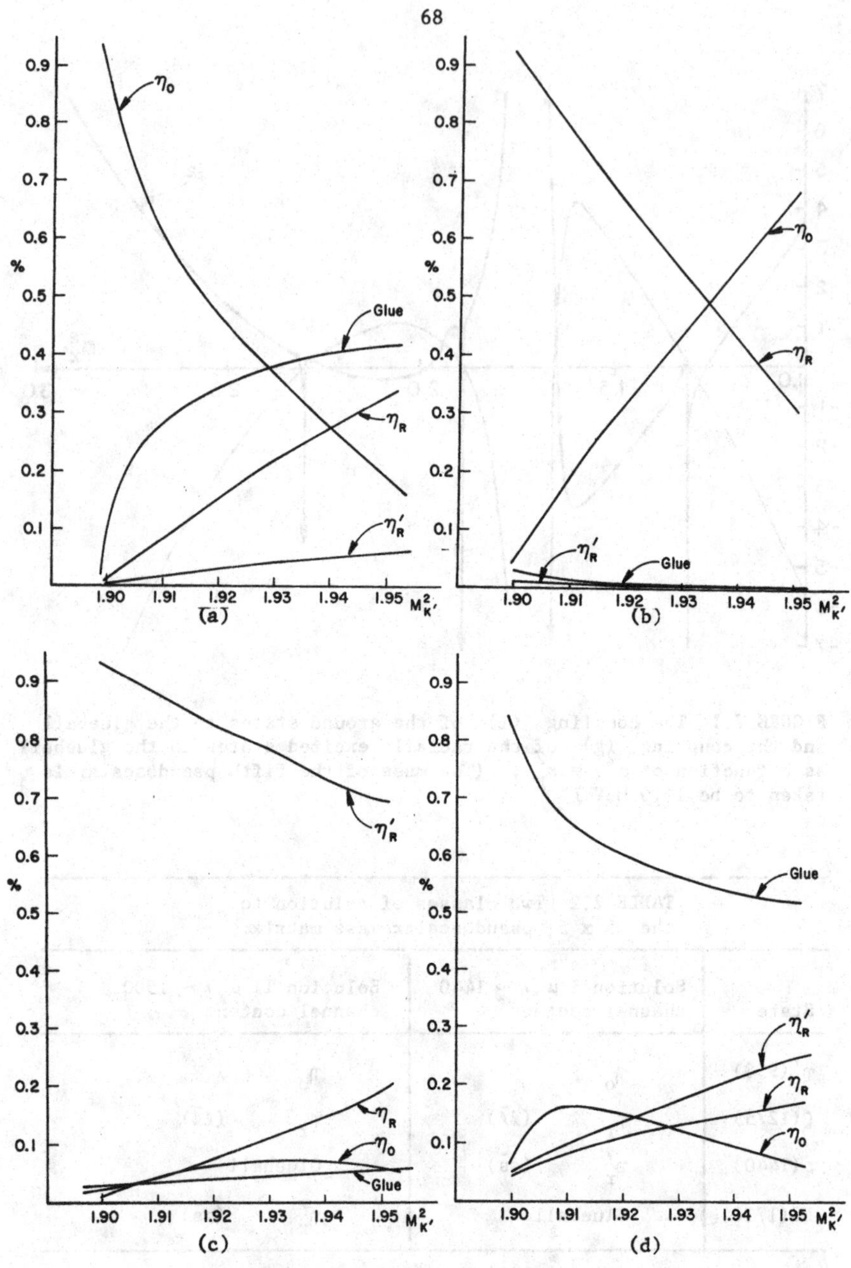

FIGURE 2.2(a-d) The channel content of the physical states for solution I.

In Figure 2.3(a)-(d) we display the corresponding information for the physical particles of solution II. In solution II (high mass side) the $\iota(1440)$ is mostly glue. [See following page.]

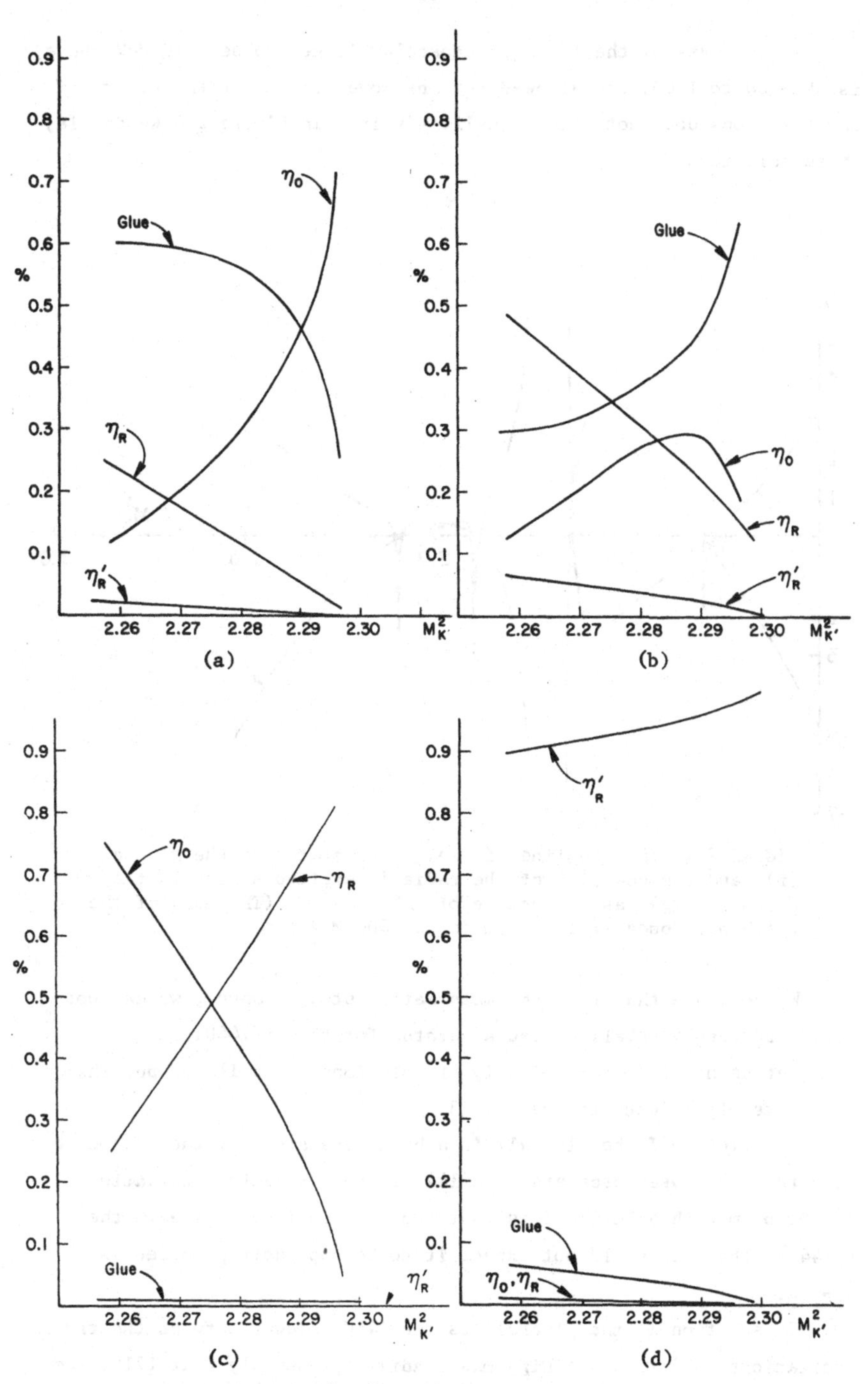

FIGURE 2.3(a-d) The channel content of the physical states for solution II.

If the mass of the fifth pseudoscalar (taken to be 1710 MeV above) is changed to 1600, the allowed regions move around but the nature of the solutions does not change qualitatively. In Figure 2.4 we display these solutions.

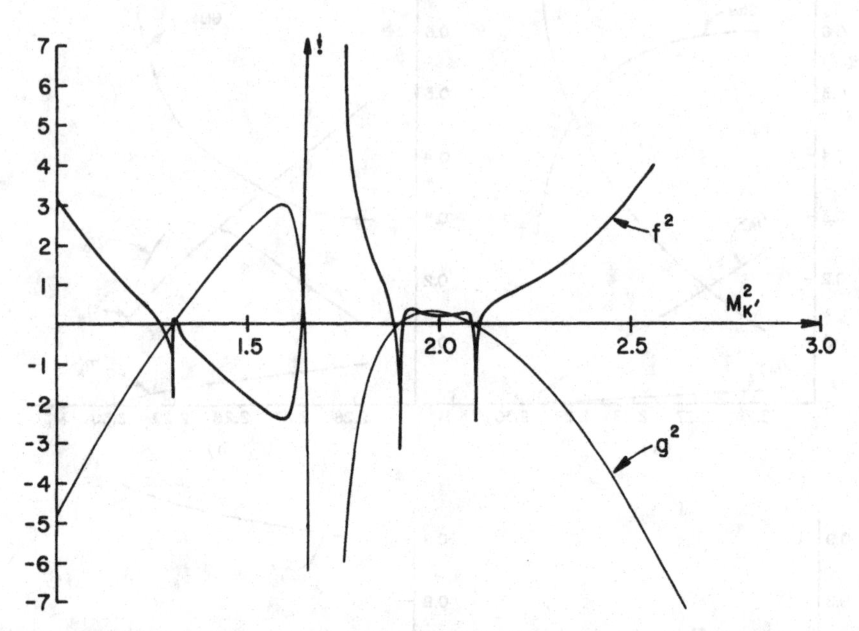

FIGURE 2.4 The coupling of the ground states to the glueball (f) and the coupling of the radially excited states to the glueball (g) as a function of $\mu^2_{K'} = \mu^2_{K'}$. (The mass of the fifth pseudoscalar is taken to be 1600 MeV.)

We conclude that from the mass matrix studies above, we can not decide between glueball or radial status for the $\iota(1440)$.

Let us now look more closely at solutions I and II and put them to the tests of the data in Eqs.(1)-(3).

In solution I the glueball is a high mass state, rather decoupled from the other pseudoscalars, and 1440 is mostly radial excitation. The problem with solution I is that there is so little glue in the $\iota(1440)$ that one would not expect it to be copiously produced in $\psi \to \gamma X$.

In solution II the glueball is the 1440, midway between the radial excitations at 1275 and a high mass radial (presumably near 1710, see below). In solution II the high mass radial ("1710") is so free of glue content that it would not be seen in $\psi \to \gamma X$; thus solution II predicts a new pseudoscalar state, above the $\iota(1440)$, which will not be seen in $\psi \to \gamma X$. (The 1710 seen in $\psi \to \gamma X$ may indeed turn out not to be a pseudoscalar. Then solution II becomes attractive.) With respect to the data in Eqs.(1)-(3), solution II does reasonably well, but has too much glue in the η' by a factor of three in the rate.

Our interpretation of solution II requires that its high mass radial excitation is not the 1710 observed in $\psi \to \gamma X$. Thus we have no reason to fix the high mass state at 1710; to this end we have varied the mass of this state and found that the character of solution II does not qualitatively change when the high mass radial is varied between 1600 and 1800 MeV.

We are currently narrowing down on solutions which best fit the data of Eqs.(1)-(3). We await determination of the spin parity and properties of the 1710 observed in $\psi \to \gamma X$. We are hopeful that new data coming from the Mk III detector at SPEAR will help us to nail down the pseudoscalars and their glueballs, if any.

This work was performed in the Department of Physics, Ohio State University, Columbus, Ohio 43210, and was supported in part by the Department of Energy.

REFERENCES

[1] D. Scharre, et al., Phys. Lett. 97B, 329 (1980).

[2] E. Bloom, Proc. 1980 SLAC Summer Inst., ed. A. Mosher, SLAC Report 239 (1981).

[3] P. Baillon, et al., Nuovo Cimento 50A, 393 (1967).

[4] D. Scharre, SLAC-PUB-2519 (1980); SLAC-PUB-2538 (1980).

[5] N. R. Stanton, et al., Phys. Lett. 92B, 353 (1980).

[6] C. Newman-Holmes, XXI International Conference on High Energy Physics, Paris, France, July 26-31, 1982.

[7] M. Chanowitz, "Glueballs and Exotic Meson Candidates", Proceedings of the APS Particles and Fields Divisional Conference (Santa Cruz, 1981); C. E. Carlson, J. J. Coyne, P. M. Fishbane, F. Gross and S. Meshkov, Phys. Lett. 99B, 353 (1981); P. M. Fishbane, D. Horn, G. Karl and S. Meshkov, "Flavor Symmetry and Glueballs", NBS Preprint, 1981.

[8] K. Ishikawa, Phys. Rev. D20, 731 (1979); Phys. Rev. Lett. 46, 978 (1981).

[9] K. A. Milton, W. F. Palmer and S. Pinsky, OSU Preprint 321. Proceedings of the XVIIth Recontre de Moriond (March 1982). W. F. Palmer and S. Pinsky, "Predictions for the $\iota(1440) \to \eta\pi\pi$";

Proceedings of the XVIIIth Recontre de Moriond (January 1983).

[10] M. Chanowitz, Proceedings of the 1981 SLAC Summer Institute (Stanford, CA 1981).

[11] J. Donoghue and H. Gomm, "Interpreting the $\iota(1440)$", University of Massachusetts Preprint UMHEP-162.

[12] P. Fishbane, G. Karl and S. Meshkov, "The $\iota(1440)$ and $\theta(1670)$ in a Mixing Model", UCLA Preprint UCLA/82/TEP/18.

[13] The exact solution of the full 5 x 5 problem with detailed numerical studies will be presented elsewhere. C. Bender, W. F. Palmer and S. Pinsky, in preparation.

GLUONIUM PHENOMENOLOGY IN MASSIVE QCD

A. Soni

Following is the outline of this talk:

1. Potential model for glueballs.[1] Its implications for the mass spectrum and for the width of gluonium states.
2. Comparison with the lattice, and bag model estimates and with the experimental candidates.
3. Deduction of the existence of a $J^{PC} = 1^{--}$ glueball through a study of $\psi(\psi')$ exclusive decays.[2,3]
4. An attempt to incorporate low-lying glueballs in hadronic symmetry.[4]
5. Summary.

1. Potential Model for Glueballs[1]

We consider the gluons to have a dynamically generated effective mass, as has been suggested by Cornwall for quite some time.[5,6] We take the intergluon potential to consist of a short distance part (V_{SD}) and a long distance part (V_{LD}). The short distance part is extracted from massive QCD and the long distance part is described by a "breakable string" which is responsible for the color screening of gluons. We take

$$V_{LD} = 2m[1 - e^{-\beta mr}] \tag{1}$$

where m is the mass of the gluon and β a screening parameter. We note that $V_{LD} \xrightarrow[r\to 0]{} 0$; $V_{LD} \xrightarrow[r\to\infty]{} 2m$ and $V_{LD} \simeq 2\beta m^2 r$ for intermediate distances. These features are compatible with the lattice studies of Bernard[7] involving adjoint Wilson loops. His parameter K_A (adjoint string tension) is given by

$$K_A = 2m^2\beta \tag{2}$$

He finds $K_A \sim (3\pm2)\ K_F$, K_F being $(0.4\ \text{GeV})^2$, $m \simeq 500$ MeV and the screening distance $r_s = 1/m\beta \simeq 1$ fm. We may also mention that the form (1) for the string potential also results in the d = 2 Schwinger model. Furthermore, Cornwall has also deduced such a potential by consideration of a condensate of vortices which massive QCD admits.[8]

The short distance part of the potential is abstracted by using massive QCD. The single gluon exchange graphs relevant for this purpose are shown in Figure 1. We stress that the on-shell amplitude is fully gauge-invariant as long as all the particles (external and internal) have the same mass. The resulting potential is a sum of Yukawa, spin-orbit, spin-spin and tensor forces when only leading relativistic corrections are retained:

$$V_{SD} = -\frac{\lambda e^{-mr}}{r}\left(\frac{2s-7m^2}{6m^2}+\frac{1}{3}S^2\right)+\frac{\pi\lambda\delta(r)}{3m^2}\left(\frac{4m^2-2s}{m^2}+\frac{5}{2}S^2\right)$$
$$-\frac{3\lambda}{2m^2}L\cdot S\,\frac{1}{r}\frac{\partial}{\partial r}\frac{e^{-mr}}{r}+\frac{\lambda}{2m^2}\left[(S\cdot\nabla)^2-\frac{1}{3}S^2\nabla^2\right]\frac{e^{-mr}}{r}, \tag{3}$$

Note that $\vec{S} = \vec{S}_1 + \vec{S}_2$ is the total spin operator and $s = (p_1+p_2)^2 = M^2$ is the glueball mass.

The above potential is singularly attractive in the 0^{++} channel due to $\delta^3(r)$ terms. We have treated this as a perturbation. We note that due to such a pathological nature of the potential and due to the mixing of the 0^{++} with the vacuum that the mass determination of the 0^{++} glueball is expected to be rather inaccurate.

The total potential

$$V = V_{SD} + V_{LD} \tag{4}$$

is thus a function of three semi-free parameters: $\lambda = 3\alpha_s$, β and m. From Bernard's work we estimate $\beta \simeq 1 \pm .7$. We also note that for $\beta \gtrsim 0.6$ the 0^{-+} glueball unbinds.

Figure 2 shows the mass spectrum of low-lying two- and three-gluon glueballs. For this calculation we have taken $\lambda = 2$ and $\beta = 0.3$. For the three-glue case the ground state (no relative angular momentum

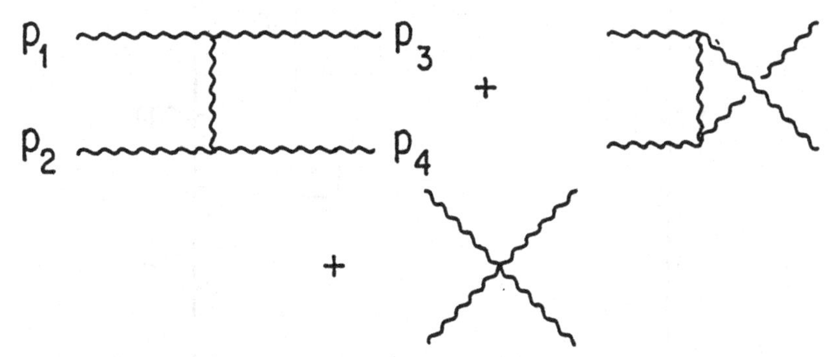

FIGURE 1 Single-particle exchange amplitude, including the seagull.

between any pair) yields three states: 0^{-+}, 1^{--}, 3^{--}. Within our approximation these states are found to be degenerate with their mass difference $\lesssim$ 50 MeV.[3] The solid and dashed lines refer to our calculations using two different ways to include relativistic corrections. The dots refer to the lattice calculations[9] and dot-cross to the bag-model calculations.[10,11] The uncertainty in each of these calculations is expected to be substantial ($\simeq \pm 25\%$) as is the case in our calculations. This is not shown (except for a few cases) to avoid overcrowding.

A few comments are in order. Bearing in mind the uncertainties in each calculation, various models roughly agree as far as the two-gluon states are concerned. The lattice and our calculations tend to give several states with mass $\simeq 1.4 \pm .4$ GeV. The bag model masses for 0^{-+} and 2^{++} ($\simeq$ 1.3 GeV) are somewhat smaller than ours. Our estimate (1.5 ± .1 GeV) for the 1^{-+} state agrees reasonably well with the lattice. This state is very important not only because it is an "oddball" (i.e. not allowed as a $q\bar{q}$ state)[12] but also because it cannot be obtained from two massless gluons and consequently it is absent in the bag model two-gluon spectrum. Schierholz et al. have recently also calculated the mass of the 0^{--} state. This state is interesting since it requires at least three gluons. We note that in our model such a state would be obtained by adding orbital angular momentum in the three-gluon system. We thus anticipate its mass to be somewhat larger than our ground state 3g mass $\simeq$ 2.4 GeV. Schierholz et al.'s result therefore tends to agree with our estimate. We note that the 3g bound states in the bag model are estimated to be at 1.4-1.8 GeV

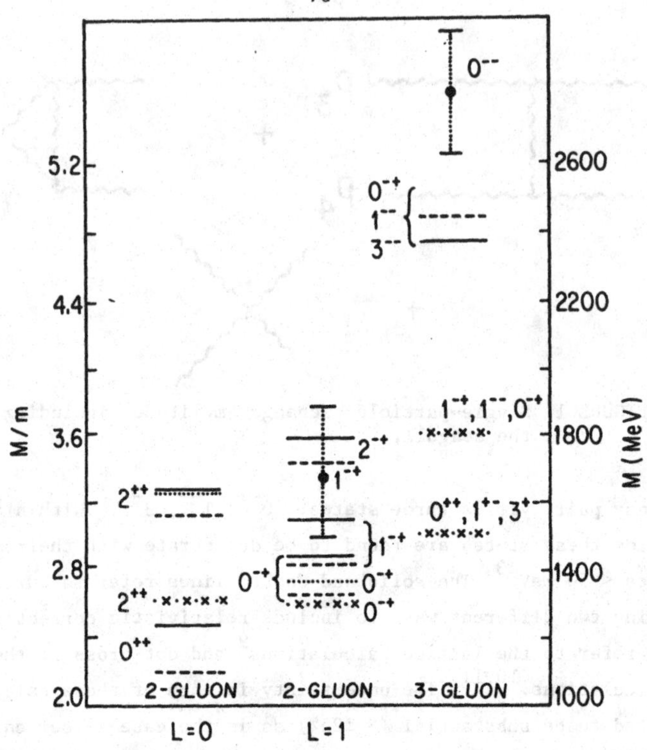

FIGURE 2 Comparison of glueball mass calculations. The solid and dashed lines refer to our work, dots refer to lattice and dot-cross to bag model calculations.

range and thus are substantially below our estimate. The 3g bound states could thus play a crucial role in discriminating between various theoretical models.

2. Glueball Widths

In the literature there have been various suggestions that the total decay widths of glueballs would follow the $\sqrt{OZI}$ rule.[13] This rule would suggest that the decay widths of glueballs being the geometric mean of the width of ordinary hadrons and that of the OZI suppressed hadrons (e.g. $c\bar{c}$ bound states) would be considerably narrower than ordinary hadrons. The $\sqrt{OZI}$ rule follows immediately when one considers the process G.B. $\rightarrow gg \rightarrow q\bar{q}$ (where q = light quark). However, in QCD the standard technique for calculating total decay widths of mesons such as ψ, η_c is via $\psi \rightarrow 3g$, $\eta_c \rightarrow 2g$. Once the gluons are produced they hadronize with unit probability. This point of view has considerable experimental support from studies of $c\bar{c}$ mesons. We have therefore suggested that the same method should be used for estimating glueball total widths i.e. by the process G.B. $\to$ gluons. Using the large N limit one then finds for the <u>total</u> decay widths M $\to$ hadrons : G.B. $\to$ hadrons : $\eta_c \to$ hadrons = 1:1:1/N. So, if one adopts this point of view one should expect glueball widths to show no suppression and to be characteristic of ordinary hadrons.

Based on the helicity suppression (a la $\pi^o \to e^+e^-$) there have been expectations in the literature that the scalar (0^{++}) and the pseudo-scalar (0^{-+}) states would exhibit extra suppression with widths being as narrow as $\simeq$ 3 MeV.[14] Due to the confinement of quarks one has to be very cautious in extrapolating the $\pi^o \to e^+e^-$ analogy to pure hadronic decays. Indeed, that such helicity suppression arguments can be very misleading is now fairly well established from non-leptonic weak decays of heavy mesons.[15] We have thus been motivated to calculate the 0^{++} glueball width in our potential model. Using the Weiskopf-Van Royen approximation[16] we have calculated the decay width for the exclusive process: G.B.$(0^{++}) \to q\bar{q}$. We find:

$$\Gamma(0^{++} \to u\bar{u} + d\bar{d} + s\bar{s}) \begin{array}{ll} \simeq 300 \text{ MeV} & \text{(constituent quark mass)} \\ \simeq \ \ 40 \text{ MeV} & \text{(current quark mass)} \end{array} \Bigg\} \qquad (5)$$

From such a calculation one can then determine the $0^{++} \to 2g$ vertex and find: $(0^{++} \to g{+}g) \simeq 600$ MeV i.e. about 50% of the mass of the 0^{++}. There are many uncertainties in such a calculation: α_s, $\psi(0)$ etc. If the scalar glueball total width is so large it is certainly going to be very difficult, if not impossible, to detect.

3. Indications of a 1^{--} Glueball in Exclusive ψ Decays[2]

Measurements[17] of the ψ, ψ' decay widths for the exclusive reactions $\psi \to X$, $\psi' \to X$ for X = $\bar{p}p$, $\bar{p}p\pi^o$, $K^+K^-\pi^+\pi^-$, $2\pi^+2\pi^-\pi^o$, $\omega\pi^+\pi^-$, $3\pi^+3\pi^-\pi^o$, $K^*\bar{K}$, $\rho\pi$ and e^+e^- are now available. From these data it is found that for each of these modes but two (namely $K^*\bar{K}$ and $\rho\pi$) the ratio: $R_{\psi'\psi} = B(\psi'{\to}X)/B(\psi{\to}X)$ is approximately constant (as is theoretically expected) $\simeq$ (12±2)%. However for the $\rho\pi$ and $K^*\bar{K}$ modes it is found that the corresponding ratio is < 1.96 and < 1.25 respectively.

We have suggested[2] that this anomalous behavior is being caused by a quantum mechanical mixing of the ground state isoscalar vector mesons (ω,ϕ,ψ) with a vector glueball (called 0). One expects the 0 to decay rather copiously to $\rho\pi$ via $0 \to \omega \to \rho\pi$ processes and to $K^*\bar{K}$ via $0 \to \phi \to K^*\bar{K}$. The experimental anomally can be explained if the contribution of this bound state (in the 3g channel) to $\rho\pi$ and $K^*\bar{K}$ is much larger than the usual 3 gluon continuum contribution to those channels. For the other 'normal' decay modes the 3-glue continuum must dominate over the glueball channel.

The fact that 0 influences ψ decays a lot more than ψ' decays suggests that the 0 is closer in mass to the ψ than to the ψ', which means $M_0 \lesssim 3.4$ GeV. Furthermore, making the simplest possible assumption for $\psi \leftrightarrow 0$ and $\psi' \leftrightarrow 0$ mixing the experiment then implies $M_0 \gtrsim 2.3$ GeV.[2] Thus

$$2.3 \text{ GeV} \lesssim M_0 \lesssim 3.4 \text{ GeV} \tag{6}$$

Such a mass for this 3g glueball could be quite compatible with our massive QCD model for glueballs where we estimate $M_0 \simeq 2.4$ GeV (see Figure 2). In contrast the bag model estimates $M_0 \simeq 1.8$ GeV.[11]

The 0 can be directly searched for as a resonance in $\rho\pi$, $K^*\bar{K}$ channels in any reaction. In particular $\psi(\psi') \to \rho\pi h$; $\psi(\psi') \to K^*\bar{K}h$ where $h = \pi\pi$ or η should be investigated. Indeed all 3g bound states with $C = \pm$ can be searched for via,[3]

$$\psi(\psi') \to h + X \ ; \quad \Upsilon(\Upsilon') \to h + X$$

where $h = \pi\pi$, η or η'. The X here is a 3g bound or continuum state. Missing mass searches of this type are obviously more difficult than $\psi \to \gamma X$ but they could be equally illuminating. Their branching ratios are expected to be a few percent. Indeed the reaction $\psi' \to \pi\pi 0$ via the sequence $\psi' \to \pi\pi\psi$ (virtual) followed by $\psi \to 0$ can be related to the observed reaction $\psi' \to \pi\pi\psi$. One then finds:[3]

$$\Gamma(\psi' \to \pi\pi 0)/\Gamma(\psi' \to \pi\pi\psi) \simeq 1\% \tag{7}$$

Experimental Candidates

Figure 3 comparés the observed experimental glueball candidates with our calculations. The best glueball candidate so far is the $\iota(1440)$.[18] The 0 is also shown. In the absence of any other viable theoretical explanation for the observed large anomalies in $\psi(\psi')$ decays the existence of the 0 has to be taken rather seriously. The glueball interpretation of the $\theta(1660)$[18] is quite tentative. For the g_T[19] and the g_S[19] seen at BNL in πp reaction too little is known at this point to make a convincing case that they are really glueballs. We do note however that if the g_T's are glueballs then in our model they are likely to be 3g bound states. This would explain their suppressed rates for production in $\psi \to \gamma X$ compared to the ι and the θ.

4. An Attempt to Incorporate Low-Lying Glueballs in Hadronic Symmetry[4]

The basic idea here is the following. Suppose one considers a phenomenological hadronic Lagrangian subject to the requirements of good high energy behavior (i.e. unitarity) and renormalizability. Then of course the resulting theory necessarily has to be a spontaneously broken gauge theory.[20] Does that necessitate the inclusion of states that can be identified as gg rather than $q\bar{q}$ states?[21]

Let us consider a chiral invariant hadronic gauge theory with the gauge group $G = SU(2)_L \times SU(2)_R$. The minimal model consists of:

$$\psi = \begin{pmatrix} n \\ p \end{pmatrix}, \quad \chi_L = \begin{pmatrix} \chi^+ \\ \chi^o \end{pmatrix}_L, \quad \chi_R = \begin{pmatrix} \chi^+ \\ \chi^o \end{pmatrix}_R, \quad M = \sigma + i\vec{\tau}\cdot\vec{\pi} .$$

Symmetry breaking takes place due to $\langle S\rangle = \langle\chi_L{}^o+\chi_R{}^o\rangle \neq 0$ and $\langle\sigma\rangle \neq 0$. The vector and axial gauge fields i.e. the ρ and the A triplets and the nucleon pick up a mass. With $\langle S\rangle = 2\sqrt{2}\, F_\pi$ and $\langle\sigma\rangle = \sqrt{2}\, F_\pi$ Weinberg sum rule $M^2{}_A/M^2{}_V = 2$ as well as other current algebra results (e.g. the correct G_A/G_V), the correct decay width for $A \to \rho\pi$ etc. are obtained. In the broken theory the particle spectrum consists of the well known hadrons n, p, ρ, A, σ and π but in addition there are two new states S and P, $P \equiv -i(\chi_L{}^o-\chi_R{}^o)/2$. These two states are glueball like states as the following arguments suggest. In the limit of no $S \leftrightarrow \sigma$ mixing one can show that $g^2{}_{S\pi\pi}/g^2{}_{\sigma\pi\pi} = (2\sqrt{2}\, F_\pi/m_\sigma)^4$

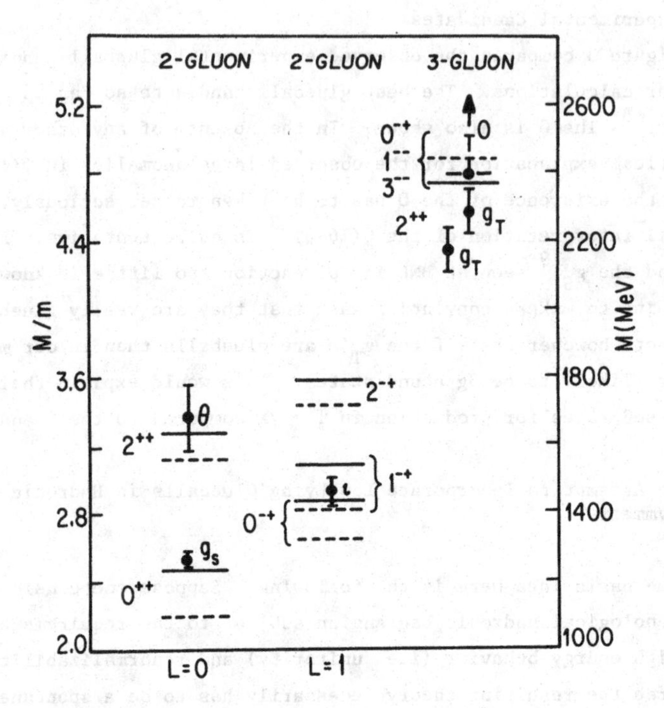

FIGURE 3 Comparison of the potential model[1] (see Figure 2) with the experimental candidates. For the ι and the θ see Reference 18; for the g_T and g_S see Reference 19, and for the O see Reference 2.

so that for m_σ very much larger than F_π the coupling of S to ππ becomes considerably weaker than that of σ to ππ. In addition, we note that nucleons have no Yukawa coupling to S, P states unlike the case for σ and π mesons. These properties are suggestive of a gg rather than a $q\bar{q}$ interpretation of S and P.

5. Summary

A potential model for glueballs based on massive QCD is discussed. It predicts two-gluon glueballs of mass ≃ 1.5 GeV and three-gluon glueballs of mass ≃ 2.4 GeV. These are compatible with lattice estimates. The bag model three-gluon glueballs are considerably lighter ≃ 1.4 to 1.8 GeV.

Comparison of ψ(ψ') exclusive decays suggests the existence of a vector glueball with mass 2.3 to 3.4 GeV. It should be searched for as a resonance in $K^*\overline{K}$ and $\rho\pi$ channels. Missing mass searches for 3g glueballs via ψ, ψ', T, T' $\to$ h + X (where h = $\pi\pi,\eta,\eta'$) are emphasized. They could be very useful in discriminating models for glueballs.

It may be interesting to incorporate low-lying glueballs in spontaneously broken gauge symmetries of hadrons.

I am very grateful to Professor J. M. Cornwall for numerous discussions. Much of the work presented in this talk was done in collaboration with him. I have also benefitted from conversations with Professor S. Meshkov. This work is supported by the National Science Foundation.

References

[1] J. M. Cornwall and A. Soni, Phys. Lett. 120B, 431 (1982).

[2] Wei-Shu Hou and A. Soni, Phys. Rev. Lett. 50, 569 (1983).

[3] Wei-Shu Hou and A. Soni, UCLA/82/TEP/14, to be published.

[4] This as well as some other material in this talk is based on work that is in progress with J. M. Cornwall and is to be published.

[5] J. M. Cornwall, in Deeper Pathways in High-Energy Physics, eds. B. Kursonoglu, A. Perlmutter and L. F. Scott (Plenum, New York, 1977).

[6] J. M. Cornwall, Phys. Rev. D26, 1453 (1982) and references therein.

[7] C. Bernard, Phys. Lett. 108B, 431 (1982); UCLA/82/TEP/20, to be published.

[8] See J. M. Cornwall, contribution to this Workshop: UCLA/83/TEP/4.

[9] K. Ishikawa, A. Sato, G. Schierholz and M. Teper, DESY preprint (1983). See also G. Schierholz, contribution to this Workshop.

[10] J. Donoghue, K. Johnson and B. Li, Phys. Lett. 99B, 416 (1981).

[11] J. Donoghue, in AIP Conference Proceedings #81, eds. C. A. Heusch and W. T. Kirk (1981), p. 97.

[12] J. J. Coyne, P. M. Fishbane, and S. Meshkov, Phys. Lett. 91B, 259 (1980).

[13] See, e.g., D. Robson, Nucl. Phys. B130, 328 (1977).

[14] C. E. Carlson, J. J. Coyne, P. M. Fishbane, and S. Meshkov, Phys. Lett. 99B, 353 (1981).

[15] M. Bander, D. Silverman, and A. Soni, Phys. Rev. Lett. 44, 7 (1980); H. Fritzsch and P. Minkowski, Phys. Lett. 90B, 455 (1980).

[16] R. Van Royen and V. F. Weisskopf, Nuovo Cimento 3, 617 (1967).

[17] G. Trilling, in Proceedings of the 21st International Conference on High Energy Physics, Paris (1982); E. Bloom, ibid; M. Franklin, Ph.D. Thesis, Stanford University (unpublished).

[18] E. Bloom, Ref. 17; C. Edwards et al., Phys. Rev. Lett. 48, 58 (1982); D. G. Coyne, in AIP Conference Proceedings #81, eds. C. A. Heusch and W. T. Kirk (1981), p. 61.

[19] A. Etkin et al., Phys. Rev. Lett. 40, 422 (1981); Phys. Rev. D25, 2446 (1982).

[20] J. M. Cornwall, D. Levin and G. Tiktopoulos, Phys. Rev. Lett. 30, 1268 (1973); Phys. Rev. D10, 1145 (1974); C. H. Llewellyn Smith, Phys. Lett. 46B, 233 (1973).

[21] J. M. Cornwall, Phys. Rev. D32, 1452 (1980).

Scalars and Pseudoscalars in Non-Linear Chiral Dynamics Based on QCD

Douglas W. McKay
and
H. J. Munczek

Department of Physics and Astronomy
University of Kansas
Lawrence, Kansas 66045
U. S. A.

The 1/N expansion of quantum chromodynamics (QCD) has recently been advocated as a guide to the construction of the effective, hadronic chiral-Lagrangian, including the consequences of the U(1) axial anomaly [1 - 4]. We extend the standard non-linear chiral Lagrangian approach and include non-linearly transforming scalar fields in addition to the usual pseudoscalars. We thereby obtain a low-energy effective Lagrangian which incorporates both the usual PCAC and U(1) anomaly constraints and the new (for non-linear Lagrangians) SU(3) breaking effects due to the scalar field vacuum expectation values. $F_K/F_\pi \neq 1$ follows naturally in our approach, for example.

Let us first summarize the transformation laws for the (3*,3) + (3,3*) fields, M and N, of our model and then describe the Lagrangian. The field M is a function only of the non-linearly transforming pseudoscalar field $\tilde{\pi} = \tilde{\pi}^a \lambda^a/2$ as in the usual model. For example, $M' = e^{i\theta} M e^{i\theta}$ under axial transformations parametrized by $\theta = \theta_a \lambda_a/2$, where we write $M = e^{2i\tilde{\pi}/F}$ and $(e^{i\tilde{\pi}/F})' = e^{i\theta} e^{i\tilde{\pi}/F} e^{-iu}$. The matrix u is a function of θ and $\tilde{\pi}$ and F is a constant with approximate value $F \simeq \sqrt{2} F_\pi = .133$ GeV. The field N is a function of the non-linearly transforming $\tilde{\pi}$ and Σ fields

($\Sigma = \Sigma^a\lambda^a/2$) where

$$N = e^{i\tilde{\pi}/F}\ \Sigma\ e^{i\tilde{\pi}/F}$$

and

$$\Sigma' = e^{iu}\ \Sigma\ e^{-iu}.$$

The scalar field Σ is assumed to have a constant part $F\sigma = \langle\Sigma\rangle$, so we write $\Sigma = F\sigma + S$. The constant matrix σ introduces the SU(3) symmetry breaking effects beyond those of the explicit (3*,3) + (3,3*) breaking term in the Lagrangian.

Turning now to the choice of Lagrangian, we write

$$L = L_{KE}(\pi,\Sigma) + V(\Sigma) + L_{SB}(\pi,\Sigma) \qquad (1)$$

The non-derivative, symmetric potential is a function of Σ alone, and the π-Σ interaction terms in the symmetric Lagrangian are all in the kinetic energy piece of the Lagrangian labeled $L_{KE}(\pi,\Sigma)$. The choice of representations and the constraints of symmetry force all of the π-Σ interactions to be of the derivative coupling type. L_{SB} is a (3*,3) + (3,3*) symmetry breaking term which involves both M and N. Elaborating the Lagrangian of Eq. (1), we have

$$L_{KE} \equiv \frac{F^2}{8}Tr(M_\mu^+ M^\mu) + \frac{aF}{16}\ Tr(M^\mu N_\mu^+ + N^\mu M_\mu^+) + \frac{1}{2}\ Tr(N_\mu^+ N^\mu), \qquad (2)$$

$$L_{SB} = F^4\,Tr\,[A(M+M^+)] + \frac{\beta}{8}F^3 Tr\,[A(N+N^+)] - \frac{1}{2}\mu_o^2(\tilde{\pi}_o)^2 \qquad (3)$$

where $\tilde{\pi}_o = \frac{1}{\sqrt{3}}\,Tr(\tilde{\pi}) = \frac{iF}{4\sqrt{3}}$ (ln det M^+ - ln det M) and A is a constant matrix proportional to the quark mass matrix. The last term in Eq. (3) breaks chiral U(1) and produces a divergence proportional to $\tilde{\pi}_o$ alone when quark mass breaking terms are absent.

A remark on the normalization of the kinetic energy term, Eq. (2), is in order here. The term bilinear in pseudoscalar fields can be written in canonical form

$$L_{KE} = \frac{1}{2}\ Tr(\partial_\mu\pi\partial^\mu\pi) + \text{-----}$$

if one defines

$$\pi^{ab} = \frac{1}{N^{ab}} \tilde{\pi}^{ab}$$

where

$$\frac{1}{N^{ab}} = \sqrt{1 + a\frac{(\sigma_a + \sigma_b)}{2} + (\sigma_a + \sigma_b)^2}$$

After mass diagonalization, the fields π^{ab} constitute the physical, tree-approximation, pseudoscalar nonet of fields. The symmetry-breaking normalization factors N^{ab} appear in the definition of the currents in terms of the fields and lead to $F_K/F_\pi \neq 1$, for example.

We are now in a position to summarize the model's parameters, the input from experiment which we use and to present results. We assume solutions where σ is proportional to A, and for amplitudes where isospin breaking is only a small correction we choose $A_{11} = A_{22} \equiv A_1$. Our parameters in these cases are then

$$a,\ \beta,\ A_1,\ A_3,\ \mu_o,\ \frac{A}{\sigma} \text{ and } F.$$

Our input choices are M_π, M_K, M_η, $M_{\eta'}$ and an $\eta - \eta'$ mixing angle θ_P which is allowed to vary from -25° to 0° (outside this range the physical pseudoscalar mass values are incompatible with the mass-matrix parametrization). Radiative decay processes are well described when $\theta_P \simeq -10°$, for example. Using $\pi^0 \to \gamma\gamma$ as input we obtain the results

$-\theta_P$	$R_\eta(.78\pm.06)$	$R_{\eta'}(1.38\pm.26)$	$R_{\psi\gamma}(.24\pm.05)$
10.5°	.95	1.42	.18

where $R_\eta \equiv |T(\eta \to \gamma\gamma) / T(\pi^0 \to \gamma\gamma)|$

$R_{\eta'} \equiv |T(\eta' \to \gamma\gamma) / T(\pi^0 \to \gamma\gamma)|$

$$R_{\psi\gamma} \equiv \frac{\Gamma(J/\psi \to \eta\gamma)}{\Gamma(J/\psi \to \eta'\gamma)} .$$

Experimental values [5] are listed in parenthesis, T means amplitude and Γ means partial width. If we take $\Gamma(\delta \to \eta\pi)$ = .05 GeV from experiment [5], we find further

$-\theta_P$	F_K/F_π	$\Gamma(\kappa \to K\pi)$	S mass scale
10.5°	1.15	0.80 GeV	1 GeV

where S mass scale means the common scalar mass value obtained when a single mass term is used for $V(\Sigma)$, Eq. (1). This acts as a consistency check on our identification of scalar fields with experimental candidates in the 1 GeV region.[5] Finally, we can determine the scalar mixing angle θ_S with input $\Gamma(S^*(980) \to \pi\pi)$ = .025 GeV and calculate the tree approximation to $\eta' \to \eta\pi\pi$. Results for two mass choices for the ε particle are shown below.

M_ε(GeV)	$\Gamma(\eta' \to \eta\pi\pi)$(180$\pm$60kev)	α(-.08$\pm$.03)
1.3	76	.04
.78	180	-.11

In summary we can say that a rather satisfactory description of low-energy spin zero physics has been obtained with a fully non-linear scalar and pseudoscalar effective chiral Lagrangian which has the QCD chiral U(1) anomaly built in.

References

[1] E. Witten, Nucl. Phys. B160, 57 (1980).

[2] C. Rosenzweig, J. Schechter and G. Trahern, Phys. Rev. D21, 3388 (1980).

[3] P. Di Vecchia and G. Veneziano, Nucl. Phys. B171, 253 (1980).

[4] P. Nath and R. Arnowitt, Phys. Rev. D23, 473 (1981).

[5] Particle Data Group, Physics Letters 111B, 22 April, 1982.

QUARK STATICS

Stephen L. Adler
The Institute for Advanced Study
Princeton, New Jersey 08540

In this talk I shall review an approach to quark binding in QCD, based on the use of effective action methods to include nonperturbative effects which are responsible for confinement. In the first part, I describe an approximation scheme which leads from the partition function of QCD to a relativistic effective action model for quark dynamics, and from this model to the leading log model for the static $q\bar{q}$ force problem. In the second part, I discuss the analysis of the leading log model, with emphasis on recent analytic investigations which give the leading behavior of the model at small and at large $q\bar{q}$ separations.

1. Approximations Leading from QCD to the Leading Log Model

Let us start from the Minkowski-space partition function for QCD,

$$
\begin{aligned}
Z(\eta,\bar{\eta}) &= \int d[\bar{\psi},\psi] \int d[A_\mu{}^a]\, \Delta_{FP}\,[A]\, e^{i\int d^4x(\mathcal{L}_{QCD}+\mathcal{L}_{gf}+\bar{\eta}\psi+\bar{\psi}\eta)} ,\\
\mathcal{L}_{QCD} &= -\frac{1}{4} F_{\mu\nu}{}^a\, F^{a\mu\nu} + \bar{\psi}(i\not{D} - m)\psi ,\\
F_{\mu\nu}{}^a &= \partial_\mu A_\nu{}^a - \partial_\nu A_\mu{}^a + g\, f^{abc}\, A_\mu{}^b A_\nu{}^c ,\\
\not{D} &= \gamma^\mu D_\mu , \quad D_\mu = \partial_\mu - i\, g\, A_\mu{}^a\, \frac{1}{2}\lambda^a \equiv D[A]_\mu ,
\end{aligned}
\tag{1}
$$

where for simplicity we consider only a single quark flavor ψ with source η. In Eq. (1), Δ_{FP} is the Faddeev-Popov compensating determinant corresponding to the background gauge-fixing Lagrangian density $\mathcal{L}_{gf}$ (relative to a background vector potential A_B),

$$\mathcal{L}_{gf}[A_B] = \frac{1}{2}(G^a)^2 \, , \qquad G^a = D[A_B]^\mu (A_\mu^{\,a} - A_{B\mu}^{\,a}) \, . \tag{2}$$

Formally evaluating the gauge-field functional integral inside the quark Grassmann integral, Eq. (1) becomes

$$Z(0,0) = \int d[\overline{\psi},\psi] \, e^{i\int d^4x \, \overline{\psi}(i\not{\partial} - m)\psi} \, e^{i\,W[J,A_B]} \tag{3a}$$

where

$$e^{i\,W[J,A_B]} = \int d[A_\mu^{\,a}]\Delta_{FP}[A,A_B] \, e^{i\int d^4x \left[-\frac{1}{4}(F_{\mu\nu}^{\,a})^2 + \mathcal{L}_{gf}[A_B] - A_\mu^a J^{\mu a}\right]} \, , \qquad J^{\mu a} = -g\,\overline{\psi}\,\gamma^\mu \,\frac{1}{2}\lambda^a\,\psi \, . \tag{3b}$$

Introducing the mean potential $\overline{A}$ (for fixed values of the quark Grassmann variables) given by

$$\overline{A}_\mu^{\,a} = -\delta W/\delta J^{\mu a} \, , \tag{4}$$

we can Legendre-transform to define the effective action functional $\Gamma[\overline{A},A_B]$,

$$W[J,A_B] = \Gamma[\overline{A},A_B] - \int d^4x \, \overline{A}_\mu^{\,a} \, J^{\mu a} \, , \tag{5a}$$

$$\frac{\delta\Gamma[\overline{A},A_B]}{\delta\,\overline{A}_\mu^{\,a}} = J^{\mu a} \, . \tag{5b}$$

In terms of Γ, we can rewrite e^{iW} as

$$e^{iW[J,A_B]} = \text{ext}_A \left\{ e^{i\Gamma[A,A_B] - i\int d^4x \, A_\mu^{\,a} \, J^{\mu a}} \right\} \, , \tag{6}$$

since the extremum in Eq. (6) determines A to have the value $\overline{A}$ given by Eq. (5b), and Eq. (5a) can then be used to invert the Legendre transform. Substituting Eq. (6) into Eq. (3), we end up with an exact expression for the QCD partition function in terms of an extremum problem [1]

$$Z(0,0) = \int d[\overline{\psi},\psi] \, \mathrm{ext}_A \left\{ e^{i\int d^4x \, \overline{\psi}(i \not{D} - m)\psi \; + i \, \Gamma[A,A_B]} \right\} . \quad (7a)$$

The fact that $Z(0,0)$ is independent of the background potential A_B used in the gauge condition is expressed through the equation

$$\frac{\delta}{\delta A_B} \; Z(0,0) = 0 \; . \quad (7b)$$

Let us now proceed to make a series of approximations, with the aim of reducing Eq. (7) to a model which can be studied numerically. (i) The first approximation is motivated by the observation that $\Gamma[A,A_B]$ is not a gauge-invariant functional of A, whereas [2]

$$\Gamma_{inv} \; [A] \equiv \Gamma[A,A] \quad (8)$$

is gauge invariant. In order to get an expression for $Z(0,0)$ in terms of Γ_{inv}, we make the mean-field approximation of pulling the extremum over A to the outside of the functional integral in Eq. (7a), giving

$$Z(0,0) \approx Z_{mf}(0,0) = \mathrm{ext}_A \int d[\overline{\psi},\psi] \; e^{i\int d^4x \; \ldots} \; , \quad (9a)$$

$$0 \approx \frac{\delta}{\delta A_B} \; Z_{mf}(0,0) \quad . \quad (9b)$$

A simple fixed-point argument [1] using Eq. (9b) shows that one can set $A_B \rightarrow A$ inside the extremum in Eq. (9a), giving the following representation for Z_{mf},

$$Z_{mf}(0,0) = \mathrm{ext}_A \left\{ \int d[\overline{\psi},\psi] \; e^{i\int d^4x \, \overline{\psi}(i \not{D} - m)\psi \, + \, i \, \Gamma_{inv}[A]} \right\} . \quad (10)$$

In the approximation of Eq. (10), the quarks move in a background gauge field $A_\mu^{\;a}$, with the dynamics of $A_\mu^{\;a}$ governed by a classical variational principle, in which $\Gamma_{inv}[A]$ is the kinetic term.

(ii) The second approximation is motivated by the fact that the extremum over A in Eq. (10), just as that in Eq. (7a), is taken over the space of all multi-bi-Grassmann valued potentials A. In order to get a model involving a number-valued background field, we make a quasi-Abelian approximation for the quark charges,

$$\lambda^a \to \hat{\lambda}^a \qquad \hat{\lambda}^{1,2,4,5,6,7} = 0 \quad , \qquad \hat{\lambda}^3 = 2\lambda^3 \; , \quad \hat{\lambda}^8 = 2\lambda^8 \quad , \tag{11a}$$

with the factor of 2 in Eq. (11a) an effective charge factor which preserves the Casimir of the charges,

$$(\hat{\lambda}^a)^2 = (\lambda^a)^2 \quad . \tag{11b}$$

It is now consistent to restrict A_μ^a to lie in the number-valued subspace

$$A_\mu^a = \hat{A}_\mu^a \quad , \quad \hat{A}_\mu^{1,2,4,5,6,7} = 0 \; , \quad \hat{A}_\mu^{3,8} = \text{number-valued} \quad . \tag{12}$$

Restoring the sources for the quarks, this gives the model described by the approximated partition function

$$\hat{Z}_{mf}(\eta,\bar{\eta}) = \text{ext}_{\hat{A}} \left\{ e^{i\,\Gamma_{inv}[\hat{A}]} \int d[\bar{\psi},\psi]\, e^{i\int d^4x[\bar{\psi}(i\,\hat{\not{D}} - m)\psi + \bar{\eta}\psi + \bar{\psi}\eta]} \right\} ,$$
$$\hat{\not{D}} = \gamma^\mu \hat{D}_\mu \quad , \quad \hat{D}_\mu = \partial_\mu - i\,g\,\hat{A}_\mu^a\,\tfrac{1}{2}\,\hat{\lambda}^a \quad . \tag{13}$$

Two remarks on the quasi-Abelian approximation are in order:

(a) In the problem with the exact charges λ^a, the flux from a quark in the fundamental 3 representation cannot be screened by octet gluons, but instead must terminate on a $\bar{3}$, which can be either an antiquark $\bar{q}$ or a diquark qq. This feature of the $q\bar{q}$ and qqq binding problems is modeled in Eqs. (11)-(13) by keeping two quasi-Abelian potentials $\hat{A}_\mu^{3,8}$, so that in the 3,8 charge space the effective charges [3] lie at the corners of an equilateral triangle centered on the origin,

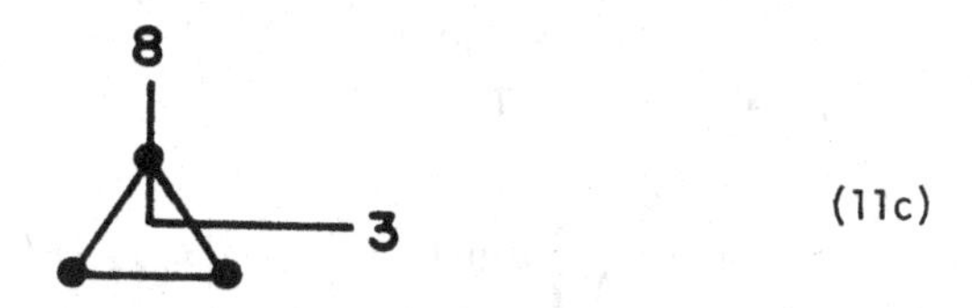

(11c)

and sum to zero for $q\bar{q}$ and qqq color singlet systems. Moreover, the equilateral triangle construction reproduces the desired relation [4]

$$V^{qqq,\ 2\ \mathrm{body}}_{\mathrm{static}}(R) = \frac{1}{2} V^{q\bar{q}}_{\mathrm{static}}(R) \quad . \tag{14}$$

Note that if the fermions in Eq. (1) were in the adjoint, rather than the fundamental representation, their charges could be screened by octet gluons, and a quasi-Abelian approximation would not correctly model the underlying physics.

(b) In the heavy quark binding problem, the quark color degrees of freedom are the only quark dynamical variables, and so the mean field approximation (i) is not needed once the quark color variables have been frozen by the quasi-Abelian approximation (ii). When both heavy and light quarks are present, the effective action formalism can be applied with a quasi-Abelian approximation made only on the heavy quark lines, and with the effects of light quark loops included exactly in $\Gamma_{\mathrm{inv}}[\hat{A}]$:

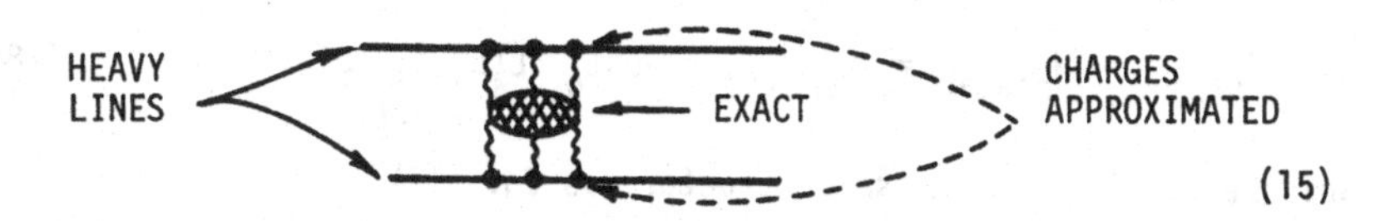

(15)

Rescaling $\hat{A} \to \hat{A}/g$, we get in the heavy quark problem

$$\lim_{T\to\infty} \hat{Z}(0,0) = \lim_{T\to\infty} e^{-i V_{static} T}$$

$$= \lim_{T\to\infty} \mathrm{ext}_{\hat{A}} \left\{ \exp \left[i\, T \int d^3x \left(\mathcal{L}_{eff} [\hat{A}/g] - \hat{A}_0^a \; J_0^a \right) \right] \right\}, \tag{16a}$$

$$+ \text{ mass terms}$$

$$\hat{J}_0^a = -\psi^\dagger \frac{1}{2} \lambda^a \psi ,$$

where we have introduced the effective Lagrangian $\mathcal{L}_{eff}$ defined by

$$\Gamma_{inv} [\hat{A}] \equiv \int d^4x \; \mathcal{L}_{eff} [\hat{A}] \; . \tag{16b}$$

Dropping mass terms, this gives the following variational principle for the heavy quark static potential,

$$V_{static} = -\mathrm{ext}_{\hat{A}} \left\{ \int d^3x \left(\mathcal{L}_{eff} [\hat{A}/g] - \hat{A}_0^a \; J_0^a \right) \right\} . \tag{17}$$

The - sign in Eq. (17) has a purely classical origin, as is seen by considering a classical dynamical system with Hamiltonian

$$H = \sum_i p_i \dot{q}_i - L(q_i, \dot{q}_i) \; . \tag{18a}$$

At a static solution of the equations of motion, where $\dot{q}_i = 0$, we have

$$V_{static} = H_{static} = -L_{static} = -\mathrm{ext}_{q_i} \; L(q_i, \dot{q}_i = 0) \; , \tag{18b}$$

which evidently has the same structure as Eq. (17).

Continuing to work with static quark sources, let us now make two further approximations, with the aim of simplifying the functional form of $\mathcal{L}_{eff}$.

(iii) The third approximation is to assume that $\mathcal{L}_{eff}$, rather than being a very complicated gauge-invariant functional of $\hat{A}/g$, has the simpler form of a local function of the field strength,

$$\mathcal{L}_{eff}\,[\hat{A}/g] \approx \mathcal{L}_{eff}\,(\hat{F}/g) \quad ,$$
$$\hat{F}_{\mu\nu} = \partial_\mu \hat{A}_\nu - \partial_\nu \hat{A}_\mu \quad . \tag{19}$$

The motivation for this approximation is that there are two classes of contributions to $\mathcal{L}_{eff}$ with qualitatively different behavior. The first are one-gluon self-energy diagrams,

(20a)

which lead to vacuum dielectric effects which are <u>linear</u> but spatially <u>nonlocal</u>. The second are multi-gluon diagrams

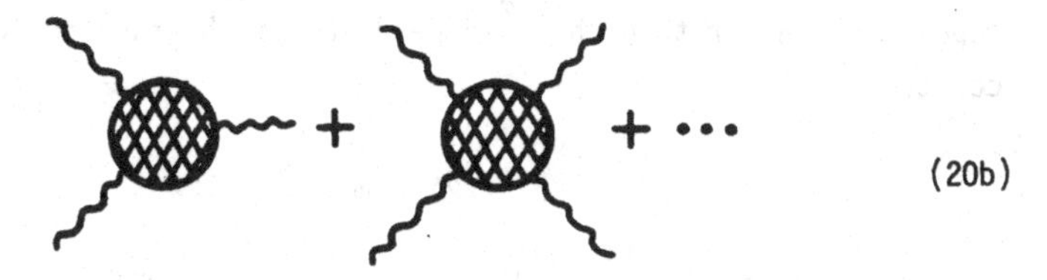

(20b)

which lead to vacuum dielectric effects which are <u>nonlinear</u> (and nonlocal). In the static quark force problem, the leading linear but nonlocal correction to the vacuum dielectric constant ε is

$$\Delta\varepsilon \sim \log Y_1 \quad , \qquad Y_1 \sim k^2 \propto \frac{1}{R^2} \quad , \tag{21a}$$

with k a typical momentum transfer and R the $q\bar{q}$ separation. The corresponding leading nonlinear correction to ε behaves as

$$\Delta\varepsilon \sim \log Y_2 \quad , \qquad Y_2 \sim |gE| \sim \frac{\alpha(R)}{R^2} \quad , \tag{21b}$$

with E a typical field strength and with $\alpha(R)$ the corresponding running coupling constant. In the infrared limit $R \to \infty$, the running coupling $\alpha(R)$ becomes infinite, so that $Y_2 \gg Y_1$ and the nonlinear, local radiative correction effects dominate the nonlocal ones [5]. Hence approximating $\mathcal{L}_{eff}$ as a local function of field strength should become reliable in the limit of large $q\bar{q}$ separations, and hence is a reasonable approximation to use in studying the confinement problem.

One gets in this way a whole class of models [6], each characterized by the particular functional form assumed for $\mathcal{L}_{eff}(\hat{F}/g)$.

(iv) As the final approximation, we keep only the classical and one-loop radiative correction terms in $\mathcal{L}_{eff}$, giving the "leading logarithm" model

$$\mathcal{L}_{eff}(\hat{F}/g) = \frac{1}{2}\frac{\mathcal{F}}{g^2}\left[1 + \frac{1}{4} b_0 g^2 \log(\mathcal{F}/\mu^4)\right]$$

$$= \frac{1}{8} b_0 \mathcal{F} \log(\mathcal{F}/e\kappa^2) \quad , \tag{22}$$

$$\mathcal{F} = -\frac{1}{2}(\partial_\mu \hat{A}_\nu^{\ a} - \partial_\nu \hat{A}_\mu^{\ a})^2 \quad , \quad \kappa^2 = \frac{\mu^4}{e} e^{-4/(b_0 g^2)} \quad .$$

This model embodies the old "infrared slavery" idea [7], but in a form in which the renormalization group is applied to the field-strength-dependence rather than the k^2-dependence of the running coupling constant,

$$\text{Eq. (22)} \iff$$

$$\mathcal{L}_{eff} \approx \frac{\mathcal{F}}{2g^2_{running}} \quad , \quad g^2_{running} = \frac{g^2(\mu^2)}{1 + \frac{1}{4} b_0 g^2(\mu^2)\log(\mathcal{F}/\mu^4)} \quad . \tag{23}$$

Factoring out the quasi-Abelian internal symmetry structure by writing

$$\hat{A}^{a0} = \hat{q}^a \phi \quad , \qquad \hat{A}^{aj} = \hat{q}^a A^j \quad ,$$

$$J_0^{\ a} = \hat{q}^a j_0 \quad , \qquad \hat{q}\cdot\hat{q} = 1 \quad , \tag{24}$$

we get the following statement of the $q\bar{q}$ static potential problem after Approximations (i) - (iv) have been made:

$$V_{static} = -\text{ext}_{\phi,\vec{A}} \int d^3x \left[\mathcal{L}_{eff}(\mathcal{F}) - \phi j_0\right] \quad ,$$

$$\mathcal{F} = \vec{E}^2 - \vec{B}^2 \quad , \qquad \vec{E} = -\vec{\nabla}\phi \quad , \qquad \vec{B} = \vec{\nabla}\times\vec{A} \quad ,$$

$$j_0 = Q[\delta^3(x - x_1) - \delta^3(x - x_2)] \quad ,$$

$$\mathcal{L}_{eff}(\mathcal{F}) = \frac{1}{8} b_0 \mathcal{F} \log (\mathcal{F}/e\kappa^2) \ , \ \kappa^2 = \frac{\mu^4}{e} e^{-4/(b_0 g^2)} \quad . \tag{25}$$

Despite the various approximations which have been made, the leading log model gives a qualitatively correct, and semiquantitatively accurate account of the $q\bar{q}$ force. We will see in Sec. 2 below that as $R = |x_1 - x_2| \to 0,\infty$ we have

$$V_{static}(R) \to \kappa Q R + O(\kappa^{\frac{1}{2}} \log (\kappa^{\frac{1}{2}} R)), R \to \infty \quad ;$$

$$V_{static}(R) \to - \frac{Q^2}{4\pi R \frac{1}{2} b_0} \left[\frac{1}{\log \left[\frac{1}{\Lambda_P^2 R^2}\right]} + O\left(\frac{\log \log}{\log^2} , \frac{1}{\log^3} \right) \right] ,$$

$$R \to 0 \ , \tag{26}$$

with $\Lambda_P = 2.52 \kappa^{\frac{1}{2}}$ for the parameter values $Q = (4/3)^{\frac{1}{2}}$, $b_0 = 9/(8\pi^2)$ appropriate to SU(3) QCD with 3 light quark flavors. Using standard [8] factors to convert from Λ_P to $\Lambda_{\overline{MS}}$, we find that within the approximations made above, $\Lambda_{\overline{MS}}$ is related to $\kappa^{\frac{1}{2}}$ by $\Lambda_{\overline{MS}} = 0.959 \kappa^{\frac{1}{2}}$. (An exact, field-theoretic determination of the relation between $\Lambda_{\overline{MS}}$ and the effective action scale mass $\kappa^{\frac{1}{2}}$ is in progress [9].) Using numerical methods to calculate V_{static} at all R [10] from Eq. (25), and making a best fit to the phenomenological heavy quark potential [11], determines the scale length $\kappa^{\frac{1}{2}}$ to be 0.23 GeV. Thus, within the leading logarithm model we find

$$\begin{aligned} (\text{STRING TENSION})^{\frac{1}{2}} &= 250 \text{ MeV} \ , \\ \Lambda_{\overline{MS}} &= 220 \text{ MeV} \ , \end{aligned} \tag{27}$$

in fairly good agreement with the respective experimental values of 400 MeV, 100 $\pm$ 50 MeV. A further nontrivial check is provided by the fact that the R-value of closest approach of the model potential to the phenomenological static potential is R = 0.68 f, which lies in the middle of the charm-upsilon region of 0.1 f to 1 f.

As an internal check on the consistency of approximations (iii) and (iv), let us verify in the leading log model that $\alpha(R) \to \infty$ as $R \to \infty$. In leading logarithm approximation, the running coupling which evolves in field strength is (for an appropriate scale mass Λ)

$$\alpha_E(R) = \frac{1}{\frac{1}{2} b_0 \log\left(\frac{\text{invariant field strength}}{\Lambda^2}\right)}$$

$$= \frac{1}{\frac{1}{2} b_0 \log\left(\frac{\alpha_E(R)}{\Lambda^2 R^2}\right)} \tag{28}$$

$$\Longrightarrow \qquad \frac{\alpha_E}{\Lambda^2 R^2} \log\left(\frac{\alpha_E}{\Lambda^2 R^2}\right) = \frac{2}{b_0 \Lambda^2 R^2}$$

Introducing the transcendental function $f(w)$ (which appears throughout the analysis of the leading log model)

$$f \log f = w \quad , \qquad f \geq 1 \quad ; \tag{29a}$$

$$f(w) \approx 1 + w + \ldots , \quad w \ll 1 , \quad f(w) \approx w/\log w \ , \quad w \gg 1 \ , \tag{29b}$$

Eq. (28) can be solved to give

$$\alpha_E(R) = \Lambda^2 R^2 \, f\left(\frac{2}{b_0 \Lambda^2 R^2}\right) \xrightarrow[R \to \infty]{} \Lambda^2 R^2 \to \infty \quad , \tag{30}$$

in agreement with the assumption made in Approximation (iii).

2. Analysis of the Leading Log Model

Let us now turn to an analytic discussion of the principal features of the leading log model.

2.1 Short-Distance Perturbation Theory [5]

Let us begin by developing Eq. (25) in a perturbation expansion for small $q\bar{q}$ separation R. Since the extremum over $\vec{A}$ lies at $\vec{A} = 0$, we are left with a nonlinear electrostatics problem

$$V_{static} = -\text{ext}_\phi \, W + \Delta V_{Coulomb} \quad ,$$

$$W = \int d^3x \left\{ \frac{1}{8} b_0 \, (\vec{\nabla}\phi)^2 \log\left[\frac{(\vec{\nabla}\phi)^2}{e\kappa^2}\right] - \phi \, j_0 \right\} ,$$

$$\Delta V_{Coulomb} = ext_\phi W_1 + ext_\phi W_2 = 2\, ext_\phi W_1 \quad ,$$
$$W_1 = W\Big|_{j_0 = Q\,\delta^3(x - x_1)} \quad , \tag{31}$$

where we have explicitly added a Coulomb counter-term to remove the infinite self-energy associated with isolated charges. Let us define a rescaled coordinate η, potential ψ and charge density ρ by

$$x = R\eta \quad , \qquad \phi = \frac{Q}{R}\,\frac{2A\,\zeta(R)}{b_0}\,\psi \ ,$$
$$j_0 = \frac{Q}{R^3}\,\rho \ , \qquad \rho = \delta(\eta - \tfrac{1}{2}\,\hat{z}) - \delta(\eta + \tfrac{1}{2}\,\hat{z}) \quad , \tag{32}$$

with A a constant to be determined and with the running coupling $\zeta(R)$ implicitly specified by the condition

$$1 = \zeta(R)\,\log\left(\frac{2Q\,\zeta(R)\,A}{b_0\,R^2\,\kappa}\right)$$

$$\Longrightarrow \zeta(R) = \frac{1}{\log f(w)} = \frac{f(w)}{w} \quad , \tag{33}$$

$$w = \frac{1}{\Lambda_P^2\,R^2} \quad , \quad \Lambda_P^2 \equiv \frac{b_0\,\kappa}{2Q\,A} \quad .$$

Substituting Eqs. (32) and (33) into Eq. (31), we can rearrange W to the form

$$W = \frac{Q^2}{R\,\frac{1}{2}\,b_0}\,\zeta(R)\,A^2 \int d^3\eta \left\{ \frac{1}{2}\,(\vec{\nabla}_\eta\psi)^2 \left[1 + \frac{1}{2}\,\zeta(R)\,\log\left((\vec{\nabla}_\eta\psi)^2/e\right)\right] - \frac{1}{A}\,\psi\rho \right\} \tag{34}$$

Since the nonlinear term in Eq. (34) is multiplied by ζ, so that W/ζ approaches the action of classical electrostatics as $\zeta \to 0$, we can expand in a perturbation series around a Coulomb solution with ζ as the small parameter,

$$\psi = \sum_{n=0}^{\infty} \left(\frac{1}{2}\zeta\right)^n \psi^{(n)} \quad ,$$
$$V_{static} = \frac{Q^2}{R\,\frac{1}{2}\,b_0}\,\zeta \sum_{n=0}^{\infty} \left(\frac{1}{2}\zeta\right)^n V^{(n)}_{static} \quad . \tag{35}$$

For the zeroth order term in the expansion we get

$$\psi^{(0)}(\eta) = \frac{1}{A}\frac{1}{4\pi}\left[\frac{1}{|\eta - \frac{1}{2}\hat{z}|} - \frac{1}{|\eta + \frac{1}{2}\hat{z}|}\right],$$

$$V^{(0)}_{\text{static}} + \Delta V^{(0)}_{\text{Coulomb}} = -\frac{1}{4\pi}, \tag{36}$$

where we note that the constant A has dropped out of the expression for the zeroth order potential. The first order term in the potential is determined entirely by $\psi^{(0)}$,

$$V^{(1)}_{\text{static}} = -A^2 \int d^3\eta \, \frac{1}{2}\left(\vec{\nabla}_\eta \psi^{(0)}\right)^2 \log\left[\frac{(\vec{\nabla}_\eta \psi^{(0)})^2}{e}\right], \tag{37}$$

and because of the logarithm the constant A now does not cancel out. Fixing A by requiring the vanishing of $V^{(1)}_{\text{static}} + \Delta V^{(1)}_{\text{Coulomb}}$ then gives

$$A^{-\frac{1}{2}} = (4\pi)^{\frac{1}{2}} e^{\frac{1}{4}} e^{T}$$

$$\Longrightarrow \frac{\Lambda_P}{\kappa^{\frac{1}{2}}} = \left(\frac{2\pi b_0}{Q}\right)^{\frac{1}{2}} e^{\frac{1}{4}} e^{T}, \tag{38}$$

where T is the convergent integral

$$T = \frac{1}{32\pi}\int d^3\eta \,(X \log X - X_1 \log X_1 - X_2 \log X_2),$$

$$X = \left(\vec{\nabla}_\eta \frac{1}{|\eta - \frac{1}{2}\hat{z}|} - \vec{\nabla}_\eta \frac{1}{|\eta + \frac{1}{2}\hat{z}|}\right)^2, \tag{39}$$

$$X_1 = \left(\vec{\nabla}_\eta \frac{1}{|\eta - \frac{1}{2}\hat{z}|}\right)^2, \quad X_2 = \left(\vec{\nabla}_\eta \frac{1}{|\eta + \frac{1}{2}\hat{z}|}\right)^2,$$

which can be evaluated numerically to give T = 0.912. Since the first-order term vanishes by construction, our final expression for the short distance perturbation expansion of the static potential reads

$$V_{static} + \Delta V_{Coulomb} = - \frac{Q^2}{4\pi R \frac{1}{2} b_0} [\zeta(R) + O(\zeta^3)] \ ,$$

$$\zeta(R) = \frac{f(w)}{w} \ , \qquad w = \frac{1}{\Lambda_p^2 R^2} \ . \qquad (40)$$

2.2 Flux-Function Formulation; Characteristics and Flux Confinement Within a Free Boundary

When expressed in terms of field-strengths, the Euler-Lagrange equations for the leading log model are

$$\vec{\nabla} \cdot \vec{D} = j_0 = Q[\delta^3(x - x_1) - \delta^3(x - x_2)] \ ,$$

$$\vec{\nabla} \times \vec{E} = 0 \ ,$$

$$\vec{D} = \varepsilon \vec{E} \ , \qquad (41)$$

$$\varepsilon = \frac{1}{4} b_0 \log (E^2/\kappa^2), \quad E = |\vec{E}| \ .$$

So far, we have introduced potentials so as to automatically satisfy the equation $\vec{\nabla} \times \vec{E} = 0$, by writing $\vec{E} = -\vec{\nabla}\phi$. In studying the large distance behavior of the leading log model, it proves useful instead to put the equations in manifestly flux-conserving form, by introducing a flux-function Φ to automatically satisfy the equation $\vec{\nabla} \cdot \vec{D} = j_0$[12]. This is done by writing

$$\vec{D} = \vec{\nabla} \times \left[\frac{\hat{\theta}}{2\pi\rho} \Phi \right] , \qquad (42)$$

where $\hat{\theta}$ is the azimuthal unit vector in cylindrical coordinates, with the charges on the z-axis. To understand the physical interpretation of Φ, consider a shell S bounded by a circle C which is rotationally symmetric around the z-axis, with radius ρ and intercept z:

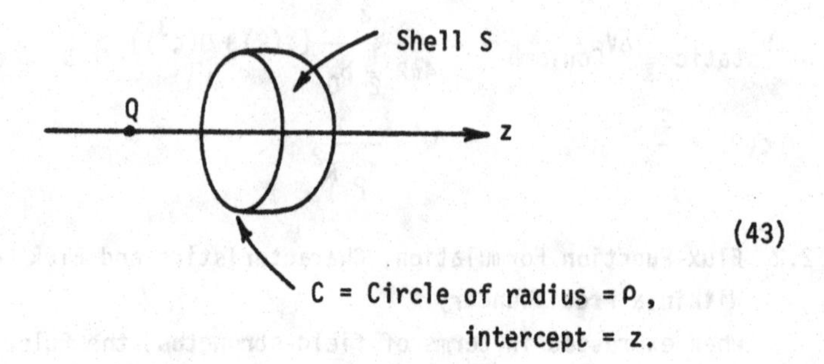

(43)

Applying Eq. (42) and Stokes's theorem, we get

FLUX THROUGH S

$$= \int_S d\vec{A} \cdot \vec{D} \tag{44}$$

$$= \int_S d\vec{A} \cdot \vec{\nabla} \times \left[\frac{\hat{\theta}}{2\pi\rho}\,\Phi\right] = \int_C d\vec{\ell} \cdot \frac{\hat{\theta}}{2\pi\rho}\,\Phi = \Phi \quad .$$

From Eq. (44), we see that Φ is discontinuous on the axis of rotation, where it takes the boundary values

$$\begin{aligned} &\Phi = 0 \qquad \rho = 0 \qquad |z| > a = \tfrac{1}{2} R = \tfrac{1}{2}\,|x_1 - x_2| \;, \\ &\Phi = Q \qquad \rho = 0 \qquad |z| < a \quad . \end{aligned} \tag{45}$$

Taken together, Eqs. (42) and (45) suffice to satisfy the equation $\vec{\nabla} \cdot \vec{D} = j_0$. The dynamical equation for Φ now comes from the equation

$$0 = \vec{\nabla} \times \vec{E} = \vec{\nabla} \times \left[\frac{\vec{D}}{\varepsilon(E(D))}\right] \quad , \tag{46}$$

which can be reduced to the form

$$\begin{aligned} &\vec{\nabla} \cdot [\sigma(\rho, |\vec{\nabla}\Phi|)\, \vec{\nabla}\Phi\,] = 0 \quad , \\ &\sigma = \frac{2\pi\kappa}{\rho|\vec{\nabla}\Phi|}\; f\!\left(\frac{|\vec{\nabla}\Phi|}{\pi\, b_0\, \kappa\rho}\right) = \frac{1}{\rho^2 \varepsilon} \quad , \end{aligned} \tag{47}$$

with f the standard transcendental function of Eq. (29). A graph of the coefficient function σ versus D has the qualitative form

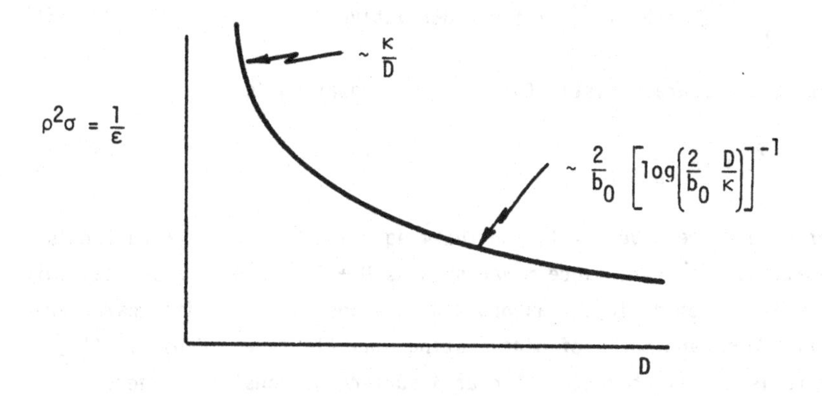

(48)

The fact that σ becomes infinite as D^{-1} as $D \to 0$ has a pronounced effect on the qualitative behavior of the differential equation for Φ [13]. This can be seen [14] by calculating the <u>characteristic form</u> of the differential equation for Φ, by collecting all second derivative terms, including those arising from differentiating the coefficient function σ. Introducing a derivative ∂_n acting normal to the level surfaces of Φ,

$$\hat{n} = \frac{\vec{\nabla}\Phi}{|\vec{\nabla}\Phi|} \quad , \qquad \partial_n = \hat{n} \cdot \vec{\nabla} \quad , \tag{49}$$

the Φ equation can be exactly rewritten as

$$\left[\partial_\rho^2 + \partial_z^2 + (\alpha - 1)\, \partial_n^2\right] \Phi - \alpha\rho^{-1}\, \partial_\rho\, \Phi = 0 \quad ,$$

$$\alpha = \frac{w}{w + f(w)} \quad , \qquad w = \frac{\partial_n \Phi}{\pi\, b_0\, \kappa\rho} = \frac{2D}{b_0 \kappa} \quad ;$$

$$\alpha = w + O(w^2) \quad , \qquad w \ll 1 \quad , \tag{50}$$

$$\alpha = 1 - \frac{1}{\log w} + O\left(\frac{\log\log w}{(\log w)^2}\right) \quad , \quad w \gg 1 \quad .$$

In terms of the tangential derivative ∂_ℓ acting parallel to the level surfaces of Φ, we have

$$\partial_\rho^2 + \partial_z^2 = \partial_\ell^2 + \partial_n^2 + \text{first derivatives} \quad , \tag{51}$$

and so the characteristic form of the Φ equation is

$$\partial_\ell^2 + \alpha\, \partial_n^2 \quad . \tag{52}$$

For D large we have $\alpha \sim 1$, and the Φ equation behaves like Laplace's equation. However, since α vanishes as $D \to 0$, the Φ equation has only a positive <u>semi</u>definite, as opposed to a positive definite characteristic form, and so is of degenerating elliptic type. Consequently, there is a real characteristic at a surface of constant Φ where $D \propto \partial_n \Phi = 0$, with $\partial_n^2 \Phi$ discontinuous across the characteristic, $\Phi \equiv 0$ outside the characteristic, and with all of the flux running from Q to -Q confined within the characteristic.

2.3 Large-Distance Structure of the Model[15]

The numerical solution [16] of the differential equation for Φ shows that in the large-R limit, the confinement domain has the following qualitative appearance:

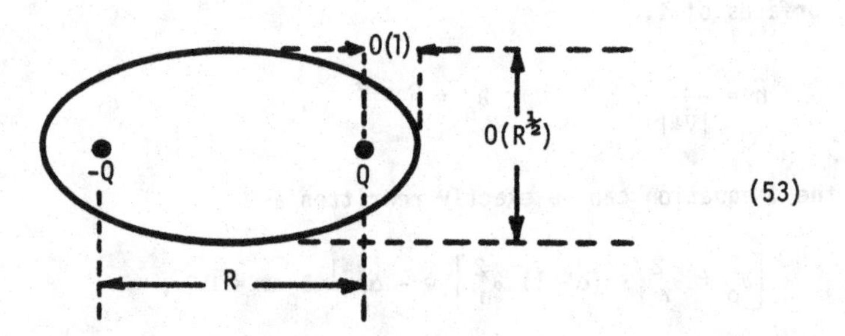

(53)

This suggests a method [15] for generating a large-R approximation for Φ and for V_{static}, as follows. To keep the formulas as simple as possible, we set $Q = 1$, $b_0 = 2$, $\kappa = 1$; general values can be recovered at the end of the calculation by simply rescaling $\pi \to \pi\, b_0\, \kappa/(2Q)$, $\Phi \to \Phi Q$. With the specialized parameter values, the differential equation to be solved is

$$\vec{\nabla} \cdot \left[\frac{1}{\rho} \frac{\vec{\nabla}\Phi}{|\vec{\nabla}\Phi|} f\left(\frac{|\vec{\nabla}\Phi|}{2\pi\rho} \right) \right] = 0 \quad . \tag{54}$$

We now make the following approximations in Eq. (54).

(i) Since in the large-R limit D is small except very near the source charges, we approximate

$$f(w) \approx 1 + w \quad . \tag{55a}$$

(ii) Motivated by Eq. (53), we substitute

$$\Phi = \Phi^{(0)} + \Phi^{(1)} + \ldots \tag{55b}$$

and do a perturbation expansion in 1/R, assuming

$$z = R\bar{z} \quad , \qquad \rho = R^{\frac{1}{2}} \bar{\rho} \; ; \qquad \bar{\rho}, \bar{z} \sim \text{unity} \quad . \tag{55c}$$

This gives as the differential equation for $\Phi^{(0)}$

$$\frac{\partial}{\partial \rho} \left\{ \frac{1}{2} \left(\frac{\partial_z \Phi^{(0)}}{\partial_\rho \Phi^{(0)}} \right)^2 + \frac{\partial_\rho \Phi^{(0)}}{2\pi\rho} \right\} + \frac{\partial}{\partial z} \left\{ - \frac{\partial_z \Phi^{(0)}}{\partial_\rho \Phi^{(0)}} \right\} = 0 \tag{56a}$$

which is solved by

$$\Phi^{(0)}(\rho, z) = \left(1 - \frac{\sqrt{\pi}}{2} \frac{a\rho^2}{a^2 - z^2} \right)^2 \quad , \quad a = \frac{1}{2} R \quad . \tag{56b}$$

The free boundary, where $\Phi^{(0)} = 0$, is an ellipsoid of revolution

$$1 = \frac{z^2}{a^2} + \frac{\sqrt{\pi}}{2} \frac{\rho^2}{a} \quad , \tag{57}$$

with semi-major axis = a and semi-minor axis $\propto a^{\frac{1}{2}}$, as expected. (Note that to leading order in 1/R, the charges lie <u>on</u> the free boundary!) The solution of Eq. (56b) can be rewritten in the form

$$\rho = \rho(\Phi^{(0)}, z) = \sqrt{\frac{2}{\sqrt{\pi}\, a}} \sqrt{a^2 - z^2} \sqrt{1 - \sqrt{\Phi^{(0)}}} \quad , \tag{58}$$

showing that the differential equation for $\rho(\Phi^{(0)}, z)$ is separable, which is how the solution was discovered. Rescaling as indicated above to general Q, b_0, κ, we have

$$\Phi^{(0)}(\rho, z) = \left[Q^{\frac{1}{2}} - \frac{1}{2}\left(\frac{\pi b_0 \kappa}{2}\right)^{\frac{1}{2}} \frac{a\rho^2}{a^2 - z^2}\right]^2 \tag{59}$$

Let us now proceed to calculate V_{static}, using the formula

$$V_{static} = \int d^3x \int_0^D dD' \; E(D') \tag{60a}$$

Substituting Eq. (55a), which implies that

$$E(D') \approx \kappa \left(1 + \frac{1}{b_0\kappa} D'\right) \tag{60b}$$

and expanding to first order in $1/R$, we get

$$\begin{aligned} V_{static} &= \int d^3x \; \kappa\left(D + \frac{1}{b_0\kappa} D^2\right) \\ &= 4\pi\kappa \int_0^{a-\varepsilon} dz \int_0^{\rho^{(0)+(1)}_{boundary}(z)} \rho \, d\rho \\ &\times \Bigg[\underbrace{-\frac{1}{2\pi\rho} \partial_\rho \left(\Phi^{(0)} + \Phi^{(1)}\right)}_{(\alpha)} \underbrace{- \frac{1}{4\pi\rho} \frac{(\partial_z\Phi^{(0)})^2}{\partial_\rho\Phi^{(0)}} + \frac{1}{b_0\kappa} \frac{1}{4\pi^2\rho^2} (\partial_\rho\Phi^{(0)})^2}_{(\beta)}\Bigg] \end{aligned} \tag{60c}$$

Since the terms labeled (α) involve $\Phi^{(1)}$ only through a total derivative, they can be evaluated by using flux conservation to give

$$V^{(\alpha)}_{static} = 2\kappa(a - \varepsilon)\left[\underbrace{\Phi^{(0)} + \Phi^{(1)}\Big|_{\rho=0}}_{Q} \; \underbrace{- \; \Phi^{(0)} + \Phi^{(1)}\Big|_{\rho=boundary}}_{0}\right] = \kappa Q R \, . \tag{60d}$$

The leading correction comes from the terms labeled (β); substituting Eq. (59) and extracting the coefficient of $\log(\kappa^{\frac{1}{2}}R)$ yields the first two terms in a large-R expansion of the static potential,

$$V_{static} \underset{R\to\infty}{=} \kappa Q R + \frac{2}{3} Q^{3/2} \sqrt{\frac{2}{\pi b_0}} \kappa^{\frac{1}{2}} \log(\kappa^{\frac{1}{2}} R) + O(1) \quad . \qquad (61)$$

This expression, as well as the small-R expansion of Eq. (40), can be used to give checks on the computer programs [16] which are used to solve the differential equation for Φ at general $q\bar{q}$ separations.

Acknowledgments

This work was supported by the U.S. Department of Energy under Grant No. DE-AC02-76ER0-2220 and the National Science Foundation under Grant No. PHY-82-17352. I wish to thank W. Dittrich for a careful reading of the manuscript.

References

[1] S.L. Adler, Phys. Lett. 110B, 302 (1982).

[2] L.F. Abbott, Nucl. Phys. B185, 189 (1981). Earlier references are given here.

[3] J. Mandula, Phys. Rev. D14, 3497 (1976).

[4] D.P. Stanley and D. Robsen, Phys. Rev. Lett. 45, 235 (1980).

[5] S.L. Adler, "Short Distance Perturbation Theory for the Leading Logarithm Models," Nucl. Phys. B (in press).

[6] H. Pagels and E. Tomboulis, Nucl. Phys. B143, 485 (1978); E. Lieb, work in progress.

[7] D.J. Gross and F. Wilczek, Phys. Rev. D8, 3633 (1973).

[8] See, e.g., A. Billoire, Phys. Lett. 104B, 472 (1981).

[9] D. Sen, work in progress.

[10] S.L. Adler and T. Piran, Phys. Lett. 117B, 91 (1982).

[11] A. Martin, Phys. Lett. 100B, 511 (1981).

[12] S.L. Adler, "The Mechanism for Confinement in Massive Quark QCD," in Proc. 5th Workshop on Current Problems in Particle Theory (Johns Hopkins University, Baltimore, 1981).

[13] Y. Nambu, Phys. Lett. 102B, 149 (1981).

[14] S.L. Adler and T. Piran, Phys. Lett. 113B, 405 (1982).

[15] T.T. Wu and H. Lehmann, work in progress.

[16] S.L. Adler and T. Piran, "Relaxation Methods for Gauge Field Equilibrium Equations," to be submitted to Rev. Mod. Phys.

Polarizable Media in Classical Gluodynamics and the SU_2 Vacuum

John P. Ralston
Dennis Sivers
Argonne National Laboratory
High Energy Physics Division
Argonne, IL 60439

Abstract

The description of a polarizable medium interacting with non-Abelian gauge fields can be formulated in a simple, gauge-invariant manner. Working with a Euclidean-space version of an SU_2 gauge-theory, we find a solution to the resulting field equations with "vacuum-like" properties:

1. $E_i^a B_i^a = 0$
2. $E_i^a E_i^a + B_i^a B_i^a$ constant
3. Area law behavior for Wilson loops formed from the classical fields.

It is possible to express the vector potential, field strength tensor and induced current in various gauges using simple analytic expressions.

Classical Continuous Media

In describing the macroscopic properties of an interacting system, it is frequently convenient to use a mathematical formulation in which the microscopic structure has been averaged. For example, the interaction of electrodynamic fields with matter is usually discussed by lumping the particles, atoms and molecules into a continuous medium with specific properties.

It is interesting to carry this approach over into a non-Abelian theory. Because non-Abelian fields themselves carry charge, it is not always possible to distinguish the interaction of the fields with "matter" and the fields with themselves through the nonlinearities in the field equations. Put another way, the gauge fields in a non-Abelian theory can contribute to the medium with which they interact.

Consider the classical Lagrangian for SU_2 gauge fields interacting with a polarizable medium in 4-dimensional Euclidean space:[1-2]

$$\mathscr{L}^{cl} = -\frac{1}{4} G^a_{\mu\nu} G^a_{\mu\nu} + \frac{1}{2} M^a_{\mu\nu} G^a_{\mu\nu} \tag{1}$$

where $G^a_{\mu\nu} = \partial_\mu A^a_\nu - \partial_\nu A^a_\mu + g\varepsilon^{abc} A^b_\mu A^c_\nu$ is the usual field strength tensor and $M^a_{\mu\nu}$ is a gauge-covariant polarization tensor which summarizes the macroscopic properties of the medium. The non-Abelian field equations are

$$(D_\mu G_{\mu\nu})^a = j^a_\nu \tag{2}$$

where $j^a_\nu = (D_\mu M_{\mu\nu})^a$ is the polarization current and $D^{ab}_\mu = \partial_\mu \delta^{ab} - g\varepsilon^{abc} A^c_\nu$ is the usual covariant derivative. The separation into "field" and "medium" is a matter of convention. Since neither $M^a_{\mu\nu}$ nor j^a_ν are gauge-invariant, we shall characterize the state of this type of classical system by the value of gauge-invariant order parameters such as $\frac{1}{4}\,{}^*G^a_{\mu\nu} G^a_{\mu\nu} = E^a_i B^a_i$ and $\frac{1}{4} G^a_{\mu\nu} G^a_{\mu\nu} = \frac{1}{2}(E^a_i E^a_i + B^a_i B^a_i)$.

In order to convince ourselves that a classical description of an extended system of non-Abelian gauge fields has possible physics content, we can use it as a framework to interpret the work on vacuum gluon condensation of Savvidy,[3], Fukuda and Kazama,[4] Nielsen and coworkers[5] or the chromostatics program of Adler.[6] Starting from an SU_2 quantum field theory, these authors take several different approaches to the quantum fluctuations in the theory. Their results can be used to motivate an effective Lagrangian

$$\mathscr{L}^{eff} = -\frac{1}{4} G^a_{\mu\nu} G^a_{\mu\nu} \left(1 + \frac{1}{4} b_0 g^2 \log\left(\frac{G^a_{\mu\nu} G^a_{\mu\nu}}{\mu^4}\right) + \ldots\right) \tag{3}$$

for the macroscopic behavior of the system. In Eq.(3) b_0 is the lowest-order coefficient in the beta function and $g^2 = g^2(\mu^2)$ is the effective coupling. In this particular example, comparison of this expression with the classical Lagrangian for a polarizable medium suggests the identification

$$\left(M^a_{\mu\nu}\right)_{vac} \cong -\frac{1}{8}\, b_0 g^2 G^a_{\mu\nu} \log\left(\frac{G^a_{\mu\nu}G^a_{\mu\nu}}{\mu^4}\right) + O(g^4) \;. \tag{4}$$

This comparison can be interpreted as saying that the quantum fluctuations in the field theory induce a condensate of gluons.[7] Macroscopic fields interact with the condensate according to the non-Abelian field equations (2).

Translationally Invariant Fields

Our basic approach is to use the classical formulation as a descriptive approximation to the nonperturbative effects in the field theory. We look for solutions to the non-Abelian Maxwell's equations, (2), characterized by $\frac{1}{4}\,{}^*G^a_{\mu\nu}G^a_{\mu\nu} = E^a_i B^a_i = 0$ and $\frac{1}{4}\,G^a_{\mu\nu}G^a_{\mu\nu} = \frac{1}{2}\,(E^a_i E^a_i + B^a_i B^a_i)$ = constant. To simplify the equations, we assume that we have, up to gauge transforms, spherical symmetry and time independence. We can then parameterize the vector potential with the ansatz

$$\begin{aligned} grA^a_4 &= \hat{r}_a f(r,t) \\ -grA^a_i &= \rho_{ia} h(r,t) + \delta^T_{ia}\beta(r,t) + \varepsilon^T_{ia}\left(\alpha(r,t) - 1\right) \end{aligned} \tag{5}$$

where $\rho_{ia} = \hat{r}_i\hat{r}_a$, $\delta^T_{ia} = (\delta_{ia} - \hat{r}_i\hat{r}_a)$ and $\varepsilon^T_{ia} = \varepsilon_{ia\ell}\hat{r}_\ell$ are tensors which mix SU_2 indices with spatial indices. The induced polarization current is parameterized

$$\begin{aligned} gj^a_4 &= \hat{r}_a q(r,t) \\ gj^a_i &= \rho_{ia} v(r,t) - \delta^T_{ia}\nu(r,t) + \varepsilon^T_{ia}\mu(r,t) \;. \end{aligned} \tag{6}$$

If we go to a radial gauge $r_i A^a_i = 0$ we can set $h = 0$ and are allowed to drop any explicit t-dependence. The field equations can then be written as a set of ordinary differential equations

$$-rf'' + \frac{2}{r} f(\alpha^2 + \beta^2) = r^2 q \tag{7a}$$

$$2(\alpha\beta' - \beta\alpha') = r^2 v \tag{7b}$$

$$\alpha'' + \frac{\alpha}{r^2}(1 - \alpha^2 - \beta^2 - f^2) = -r\mu \tag{7c}$$

$$\beta'' + \frac{\beta}{r^2}(1 - \alpha^2 - \beta^2 - f^2) = -r\nu \tag{7d}$$

$$rv' + 2v + 2(\nu\alpha - \mu\beta) = 0 \tag{7e}$$

where the primes denote radial derivatives. This system can be simplified further by imposing the constraint $E_i^a B_i^a = 0$. Using the radial gauge condition the field strength tensor can be written

$$gr^2 E_i^a = \rho_{ia}(f - rf') - \delta_{ia}^T \alpha f + \epsilon_{ia}^T \beta f \tag{8a}$$

$$gr^2 B_i^a = \rho_{ia}(1 - \alpha^2 - \beta^2) - \delta_{ia}^T r\alpha' + \epsilon_{ia}^T r\beta' \tag{8b}$$

and the constraint $E_i^a B_i^a = 0$ yields

$$E_i^a B_i^a = -\frac{1}{r^2}\left(\frac{f}{r}(1 - \alpha^2 - \beta^2)\right)' = 0 . \tag{9}$$

This can be integrated to give

$$\frac{f}{r}(1 - \alpha^2 - \beta^2) = \lambda , \tag{10}$$

and attention to the conservation of topological charge gives $\lambda = 0$.[2] From (8a) and (10) we see that the existence of an electric field implies $\alpha^2 + \beta^2 = 1$. This means we can write

$$\alpha = \cos(\omega(r))$$
$$\beta = \sin(\omega(r)) \tag{11}$$

and the non-Abelian Maxwell's equations can be simplified to

$$-rf'' + \frac{2}{r} f = r^2 q \tag{12a}$$

$$2\omega' = r^2 v \tag{12b}$$

$$\omega'' = r(\mu \sin\omega - \nu \cos\omega) \tag{12c}$$

$$\omega'^2 + \frac{f^2}{r^2} = r(\mu \cos\omega + \nu \sin\omega) \ . \tag{12d}$$

We next have to consider the behavior of $G^a_{\mu\nu}G^a_{\mu\nu} = 2(E^a_i E^a_i + B^a_i B^a_i)$. Using (8) and (10) we see that we can write

$$g^2 E^a_i E^a_i = [(\frac{f}{r})']^2 + 2(\frac{f}{r^2})^2 \tag{13a}$$

$$g^2 B^a_i B^a_i = 2(\frac{\omega'}{r})^2 \ . \tag{13b}$$

Requiring the medium to be uniform when subjected to Euclidean boosts gives

$$E^a_i E^a_i = B^a_i B^a_i = \text{constant}$$

which yields with (13),

$$f = \sigma r^2 \tag{14a}$$

$$\omega = \frac{1}{2} \ell r^2 + \omega_0 \tag{14b}$$

with $2\ell^2 = 3\sigma^2$ and where ω_0 is an overall phase which represents a residual gauge freedom; we set $\omega_0 = 0$ here.

It is interesting to note that (14a) is a solution to (12a) with $q(r) = 0$. There is no static color electric charge induced in the medium by this type of spatially uniform gluon condensation. This is consistent with the hypothesis put forward by Mandelstam[8] that the gluon condensate can be understood as a coherent superposition of magnetic charges.

The results can be summarized by the set of (radial gauge) expressions:

$$A_4^a = \frac{\sigma}{g} r \hat{r}_a$$

$$A_i^a = \frac{-1}{gr} \left(\delta_{ia}^T \sin\left(\frac{\ell r^2}{2}\right) + \epsilon_{ia}^T \left(\cos\left(\frac{\ell r^2}{2}\right) - 1\right)\right) \tag{15a}$$

$$j_4^a = 0$$

$$j_i^a = \frac{2\ell}{rg} \rho_{ia} - \frac{\ell}{r\sigma} E_i^a - \frac{(\ell^2 + \sigma^2) r}{\ell} B_i^a \tag{15b}$$

with

$$-E_i^a = \frac{\sigma}{g} \left(\rho_{ia} + \delta_{ia}^T \cos\left(\frac{\ell r^2}{2}\right) - \epsilon_{ia}^T \sin\left(\frac{\ell r^2}{2}\right)\right)$$

$$B_i^a = \frac{\ell}{g} \left(\delta_{ia}^T \sin\left(\frac{\ell r^2}{2}\right) + \epsilon_{ia}^T \cos\left(\frac{\ell r^2}{2}\right)\right) . \tag{15c}$$

A simple picture in this gauge is that the E^a and B^a fields have fixed length and rotate with period, $\omega = \ell r^2/2$, keeping a fixed right angle between them.

We can understand more about the properties of the medium by looking at

$$W(s) = g \oint_s \tau^a A_\mu^a(x) dx_\mu(s) . \tag{16}$$

Forming a loop (s) corresponding to the radial separation of hypothetical test charges which separate, propagate for a long time (T) and then come together, one finds

$$\begin{aligned} W(r_2, r_1, T) &= g\left[A_4^a(r_2) - A_4^a(r_1)\right]\hat{r}_a T \\ &= \sigma(r_2 - r_1)T \\ &= \sigma(\text{Area}) . \end{aligned} \tag{17}$$

We see that the polarization current in the medium opposes the separation of color charges. We cannot, of course, discuss a property such as "confinement" without the full mechanism of a quantum field theory and pair creation. However, this does allow us to identify the parameter σ with an "effective" string tension.

We can transform our solutions to other gauges using the usual formalism. Transformations are particularly convenient when the gauge function can be written $\Omega(\vec{x},t) = \exp(i\tau_a \hat{r}_a \theta(r,t)/2)$. For example we can transform to an $A_4^a = 0$ gauge to find

$$
\begin{aligned}
-gA_i^a(r,t) &= \frac{1}{r}\left(\sigma rt\ \rho_{ia} + \delta_{ia}^T \sin(\sigma rt) + \varepsilon_{ia}^T\left(1 - \cos(\sigma rt)\right)\right) \\
E_i^a &= -\frac{\sigma}{g}\left(\rho_{ia} + \delta_{ia}^T \cos(\sigma rt) - \varepsilon_{ia}^T \sin(\sigma rt)\right) \\
B_i^a &= \frac{\ell}{g}\left(\delta_{ia}^T \sin(\sigma rt) + \varepsilon_{ia}^T \cos(\sigma rt)\right) .
\end{aligned}
\tag{18}
$$

Simple expressions can be found in many other convenient gauges[2]. We can also calculate in terms of σ higher-order gauge-invariant order parameters such as

$$
\begin{aligned}
\varepsilon_{ijk}\varepsilon^{abc} E_i^a E_j^b B_k^c &= 0 \\
\varepsilon_{ijk}\varepsilon^{abc} E_i^a B_j^b B_k^c &= \frac{3}{2}\frac{\sigma^3}{g^3} .
\end{aligned}
\tag{19}
$$

More details concerning this topic can be found in Reference 2. The possibility that other types of extended non-Abelian systems can be analyzed using this classical formulation seems worth consideration.

References

[1] D. Sivers, Phys. Rev. D27, 947 (1983).

[2] J. Ralston and D. Sivers, ANL-HEP-PR-83-19.

[3] G. K. Savvidy, Phys. Lett. 71B, 133 (1977)

[4] R. Fakuda and Y. Kazama, Phys. Rev. Lett. 45, 1142 (1980).

[5] H. Nielsen and P. Olesen, Nucl. Phys. B144, 376 (1978); NBI-HE-79-45 (unpublished).

[6] S. L. Adler, Phys. Rev. D23, 2905 (1981).

[7] M. A. Shifman, A. I. Vainstein and V. I. Zakharov, Nucl. Phys. B147, 385 (1979); ibid. 448 (1979); ibid. 519 (1979).

[8] S. Mandelstam, Phys. Rep. 23C, 245 (1976).

EFFECTIVE QUARK PROPAGATOR AND $Q\bar{Q}$ STATES IN QCD

H. Munczek and A. M. Nemirovsky

Department of Physics and Astronomy
The University of Kansas
Lawrence, Kansas 66045

1. Introduction

We discuss in this contribution the role of the quark propagator in an approximation scheme [1] to calculate masses and wavefunctions of the $Q\bar{Q}$ states in quantum chromodynamics. The scheme is based on the Bethe-Salpeter equation for the $Q\bar{Q}$ amplitude solved together with the Schwinger-Dyson equation for the quark propagator.

Approaches similar to the one we discuss here have been used previously to study QCD in 1+1 dimensions [2], in a model with spinless quarks [3] and also in the fully relativistic case [4-7]. The scheme we present provides a separation of the mass of the boson into a main part and corrections due to finer details of the interaction between quarks. This separation is compatible with PCAC, and also illuminates the relationship between current algebra and constituent quark mass parameters.

The quark propagator, solution of our S-D equation, has no poles. In addition, for massless quarks, it has a scalar piece that breaks the γ_5 invariance. Then, massless pions, the Nambu-Goldstone bosons of the broken chiral symmetry, appear together with massive vector mesons. With the quark propagator and the B.S. equation for the bound states, we obtain a good fit to both the pseudoscalar and vector ground state masses for all flavor combinations. Our only input are the (current algebra) quark masses and one constant of dimension of mass. Also, in the same approximation as for color singlets, we find that there are no solutions of the bound state equations for colored mesons.

2. The Bound State Equation and the Equation for the Fermion Propagator.

We start with the momentum space Bethe-Salpeter equation for the color singlet $Q\bar{Q}$ amplitude ϕ (p,k)

$$S^{-1}(k+\alpha p)\phi(p,k)S^{-1}(k-\beta p) = \frac{4}{3}ig^2 \int \frac{dq}{(2\pi)^4} G_{\mu\nu}(q)\gamma^\mu \phi(p,k-q)\gamma^\mu \qquad (1)$$

where ϕ is a matrix in both, spinor and flavor indices. p and k are respectively the total and relative momentum of the $Q\bar{Q}$ system. $G_{\mu\nu}(q)$ is the gluon propagator, α and β are two positive quantities, such that $\alpha + \beta = 1$. The flavor diagonal matrix S is an approximation to the quark propagator and it satisfies the Schwinger-Dyson equation

$$S(k) = [\not{k} - M - \frac{4}{3}ig^2 \int \frac{dq}{(2\pi)^4} G_{\mu\nu}(q)\gamma^\mu S(k-q)\gamma^\nu]^{-1}, \qquad (2)$$

where M is the bare, or current algebra, quark mass matrix.

Equations (1) and (2) are approximations to the exact equations since they contain the vertex function only to zeroth order. They still maintain (when M=0) the original invariance under chiral global transformations. It has been shown [8] that when this invariance is spontaneously broken, i.e., when $\{S,\gamma_5\} \neq 0$, the exact equations imply the existance of massless pseudoscalars. We expect then that, as long as our approximation maintain chiral invariance and the consistency of the equations, this feature will persist. As we show later, this is indeed the case.

We now introduce the following decomposition of the gluon propagator

$$i \frac{4}{3} g^2 \frac{G^{\mu\nu}}{(2\pi)^4} = -[\frac{1}{4} \eta^2 \delta^4(q) + V(q)] g^{\mu\nu}, \qquad (3)$$

where η^2 has units of mass-squared. In configuration space the first term on the right gives by itself, a constant gluon propagator. The second term determines the behavior at the origin and at large distances. We shall assume that the constant represents the value of the potential in the region of average quark-antiquark separation, and we propose to use it as zeroth-order approximation to the potential in a scheme to solve both the bound state equation and the equation for the fermion propagator, Eqs. (1) and (2).

Next we define two operators, $L(p,0)$ and $K(p,k)$, that act on all indices of the matrix ϕ

$$S^{-1}(k+\alpha p)\, \phi\, (p,k) S^{-1}\, (k-\beta p) \equiv [L(p,0) + K(p,k)]\, \phi\, (p,k). \tag{4}$$

Using Eqs. (3) and (4) we can express Eq. (1) in the following form

$$\begin{aligned} &K(p,k)\phi(p,k) + \int V(q)\, \gamma_\mu\, \phi\, (p,k-q)\gamma^\mu\, dq \\ &+ [L(p,0)\, \phi\, (p,k) + \frac{1}{4}\, \eta^2 \gamma_\mu\, \phi\, (p,k)\gamma^\mu] = 0 \quad . \end{aligned} \tag{5}$$

Eq. (5) is an eigenvalue equation for the operator $K + V$. In general such an equation will have solutions only for specific values of p^2, which give the squared mass eigenvalues. For these values of p^2 the bracket in the equation is an eigenvalue matrix acting on $\phi(p,k)$. Since K is a function of the relative kinetic energy and V is assumed to give either the very short, or long distance behavior, we approximate the ground state by taking $K + V = 0$. Then the bound state equation for the mass eigenvalue reduces to the condition that the bracket in Eq. (5) vanishes, i.e.,

$$L(p,0)\, \phi\, (p,k) + \frac{1}{4}\, \eta^2\, \gamma_\mu\, \phi\, (p,k)\, \gamma^\mu = 0 \tag{6}$$

Eq. (6) can be written explicitly as

$$S^{-1}\, (\alpha p)\, \phi\, (p,k)\, S^{-1}\, (-\beta p) + \frac{1}{4}\, \eta^2\, \gamma_\mu\, \phi\, (p,k)\, \gamma^\mu = 0 . \tag{7}$$

Consistently S should be taken to be the solution of Eq. (2) with $V = 0$,

$$(\not{k} - M + \frac{1}{4}\, \eta^2 \gamma_\mu\, S\, (k) \gamma^\mu)\, S(k) = 1 \quad . \tag{8}$$

3. The Quark Propagator

To solve Eq. (8) we write S as a flavor diagonal matrix

$$S(k) = - \frac{2}{\eta^2}\, [\not{k}\sigma_V(k) + \frac{1}{2}\, \sigma_S(k)] \quad . \tag{9}$$

Inserting this expression into Eq. (8) we obtain two coupled equations for σ_s and σ_v.

$$(1+\sigma_v)\,\sigma_s - 2\,(M+\sigma_s)\,\sigma_v = 0 \quad , \tag{10a}$$

$$2k^2\,\sigma_v(1+\sigma_v) - (M+\sigma_s)\,\sigma_s = -\,\eta^2 \quad , \tag{10b}$$

which, for $M \neq 0$, can be recast in the following form

$$\sigma_v = \frac{\sigma_s}{2M+\sigma_s} \quad , \tag{11a}$$

$$k^2 = \frac{1}{4}\,(2M+\sigma_s)^2\,[1 - \frac{\eta^2}{\sigma_s(M+\sigma_s)}] \quad . \tag{11b}$$

From now on we take η^2 to be real and positive. For each value of k^2 there are four values of σ_s. The solution chosen in Ref. 1 has the property that as σ_s ranges monotonically and continuously over the real positive axis, k^2 ranges over the full real axis. Also, all derivatives of σ_s are continuous on the real k^2 axis. Moreover, we see from Eq. (11a) that σ_v has the same properties as σ_s. Then, it follows that the fermion propagator has no singularities for real k^2. In addition the homogeneous equation $S^{-1}\,\psi = 0$ has no solution since our σ_s can take only positive real values. This property seems to be a reasonable necessary criterion for confinement [9].

As $M \to 0$, and for real k^2, σ_s and σ_v tend to $\sigma_s = (\eta^2+4k^2)^{\frac{1}{2}}$, $\sigma_v=1$; for $\eta^2 + 4k^2 > 0$. For $\eta^2 + 4k^2 < 0$, we have instead, $\sigma_s = 0$, $\sigma_v = \frac{1}{2}\,(1 - \frac{2\eta^2}{k^2})^{\frac{1}{2}} - \frac{1}{2}$. This solution can be obtained from Eqs. (10a,b) when $M = 0$ and it corresponds to a spontaneous breaking of the global chiral symmetry displayed by Eq. (1) when $M = 0$.

4. Solutions of the Bound State Equation.

We now discuss the solutions to Eq. (7) which have quantum numbers $J^{PC} = 0^{++}, 0^{-+}, 1^{--}, 1^{++}$. For our ground state, the most general structure [10] of these wavefunctions is

$$\phi = S + S_1\,\not{P}, \tag{12a}$$

$$\phi_p = \gamma_5\,(P + P_1\,\not{P}), \tag{12b}$$

$$\phi_V = [V(\gamma_\mu - \not{p}p_\mu/p^2) + V_1\, p^\nu\sigma_{\mu\nu}]\, \varepsilon^\mu, \tag{12c}$$

$$\phi_A = [A\, \gamma_5(\gamma_\mu - \not{p}p_\mu/p^2) + A_1\, \gamma_5\, p^\nu\sigma_{\mu\nu}]\varepsilon^\mu\ , \tag{12d}$$

where ε^μ is an external vector. In these expressions S, P, etc. are scalar functions of k and p. They are also matrices in flavor indices, i.e., $S = S_a{}^b$.

By substituting each of the forms given by Eqs. (12) into Eq. (7), and projecting the appropriate Dirac matrices one obtains two coupled homogeneous equations for S, S_1; P, P_1; etc. Since we are interested in solutions other than the trivial one, we require that, for a given spin-parity and flavor combination, the determinant $D(p^2)$ of the coefficients vanishes. Studying the condition $D(p^2) = 0$ for 0^{++}, 0^{-+}, 1^{--} and 1^{++} mesons of different flavor combinations we conclude

a) The equations $D_S = 0$ and $D_A = 0$ have no solution, for any spin-parity channel and flavor combination.

b) The equations $D_P = 0$ and $D_V = 0$ have only one real solution, for a given spin-parity channel and flavor combination. The p^2 solution of $D(p^2) = 0$ is always positive (timelike), consistent with p^2 identified as the mass-squared of the meson.

c) For massless quarks, the solutions of the equations $D_P = 0$ and $D_V = 0$ are $p^2 = 0$ and $p^2 = \eta^2/2$ respectively. As expected we obtain a massless pseudoscalar. If the massive vector meson is identified with the rho meson, we estimate $\eta^2 \approx 1\ \mathrm{GeV}^2$.

d) For pseudoscalar mesons, composite of light quark and antiquark, both of mass m ($m \ll \eta$) we can obtain $p^2 = 3/2\, m\eta + O(m^2)$ in agreement with the general PCAC analysis. [11].

e) For mesons, composite of heavy quark and antiquark, both of mass m ($m \gg \eta$) we obtain $p^2 = 4m^2 + O(m\eta)$ as solutions of both $D_P = 0$ and $D_V = 0$ (the corrections are, of course, different for pseudoscalar and vector mesons).

f) The results of the numerical calculation of the mass spectrum, reported in Ref. [1] show good agreement with experiment. The parameters, quark masses and η^2, were chosen by fitting the pion and the diagonal vector mesons to their experimental values. We obtained $m_u = m_d = 12$ MeV; $\eta^2 = 1.14\ \mathrm{GeV}^2$; $m_s = 175$ MeV, $m_c = 1.37$ GeV, $m_b = 4.65$ GeV.

g) Finally, in the color octet channel, we found that there are no solutions of $D(p^2) = 0$ for any spin-parity and flavor combination.

5. Conclusions

Our results for the quark propagator and for the ground state spectrum show that our approximation scheme provides a promising framework for the relativistic calculation of $Q\bar{Q}$ states. Until more is known about the nonconstant part of the gluon propagator in equation (3) one will have to rely on phenomenological forms for $V(q)$ in order to account for $L \geq 1$ and radially excited states [12].

An essential ingredient in our calculation is the quark propagator. In the $V = 0$ approximation it has properties which can be interpreted as indicating confinement. When used in a form consistent with the Bethe-Salpeter equation, it preserves the chiral properties of the exact equations and leads to the existence of both Nambu-Goldstone pions and massive vector mesons with the same light quark content.

REFERENCES

[1] H. Munczek and A. M. Nemirovsky, University of Kansas preprint. To be published in Phys. Rev. D.

[2] G. 't Hooft, Nucl. Phys. B75, 461 (1974); C. Callan, N. Coote, and D. Gross, Phys. Rev. D13, 1649 (1976).

[3] C. Alabiso and G. Schierholz, Nucl. Phys. B110, 81 (1976); B110 93 (1976).

[4] H. Pagels, Phys. Rev. D14, 2747 (1976); D15, 2991 (1977).

[5] A. Swift and F. Roig, Phys. Rev. D18, 1306 (1978).

[6] S. Mandelstam, Phys. Rev. D20, 3223 (1979).

[7] U. Bar-Gadda, Nucl. Phys. B163, 312 (1980).

[8] K. Lane, Phys. Rev. D10, 2605 (1974).

[9] Quark propagators interpreted as indicating confinement have been discussed in Refs. 4-7 above, and also by J. M. Cornwall, Phys. Rev. D22, 1452 (1980); and by J. S. Ball and F. Zachariasen, Phys. Lett. 106B, 133 (1981).

[10] C. H. Llewellyn Smith, Ann. Phys. (N.Y.) 53, 521 (1969).

[11] M. Gell-Mann, R. J. Oakes and B. Renner, Phys. Rev. 175, 2195 (1968).

[12] A calculation using Eq. (5), but with V(q) = 0, shows that one can obtain solutions for the L = 1, scalar and axial vector states; A. M. Nemirovsky, University of Kansas preprint.

FINDING DYNAMICAL MASSES IN CONTINUUM QCD

John M. Cornwall

1. Introduction

In QCD without quarks, there are many dimensionful parameters, all of them scaling as various powers of the renormalization group mass Λ_{RG}: the fundamental (quark) string tension K_F, the adjoint (gluon) string tension K_A, vacuum expectation values like $\langle g^2 \sum^a (G_{\mu\nu}{}^a)^2 \rangle \equiv \langle g^2 G\cdot G \rangle$, the closely-related vacuum energy density Ω_{VAC}, and $\langle S \rangle$, where $S(x)$ is the field operator for the $J^{PC} = 0^{++}$ glueball of lowest mass. In addition, of course, there are the masses of all the glueballs.

In this paper, we review and extend earlier work [1,2,3] which argues that the simplest way to understand and to calculate these numbers is by recognizing that the strong interactions of QCD generate a dynamical gluon mass m, in much the same way that the photon of the Schwinger model (massless d = 2 QED) becomes massive. In both cases local gauge invariance is completely preserved; there is no breaking of color symmetry. There are, however, very important differences: QCD gluons are not observable as physical particles, hence neither is their mass; and this mass vanishes at large momentum or short distances. That is, the mass is dynamical in origin, and plays no fundamental role in the Lagrangian, which has no scale of mass. (Ref. [4] contains many reprints on dynamical mass generation; see also [9].) Because isolated gluons do not exist, it is not clear, at first, precisely how to ascribe any mass (even zero!) to gluons. There is, as we discuss later, a definite prescription for extracting the mass as a pole in a special Green's function which is superficially related to the gluon propagator. However, the new Green's function, quite unlike the propagator, is gauge-invariant, and replaces the useless propagator as an elemental description of part of the force law due to gluon exchange. At a more advanced stage in the description, the pole is shielded by confining forces which themselves arise as natural consequences of the gluon mass, or force-law pole. It should cause us no concern that this mass is not directly observable, any more than we should worry that Λ_{RG} can only be compared with experimental data after laborious computations.

Similarly, many difficult calculations are needed to express the dimensional parameters of QCD in terms of m, or to express m in terms of Λ_{RG}. Important as these calculations are, we relegate them (with one exception) to secondary importance in this report for several reasons: there is not space here to do them justice; some of the work bv the author and A. Soni [5] on glueballs will be presented elsewhere in these Proceedings; and finally it is more important to understand the qualitative flow from mass to confinement without having to worry about technical detail. Thus in Section 2 we give the qualitative arguments, in Section 3 we report for the first time a fairly simple-minded calculation of the (breakable) string potential between gluons, and finally in Section 4 we review briefly some of the technical arguments underlying the generation of gluon mass, and which have led [3] to the estimate $m \simeq 2\Lambda_{mom} \simeq 400$ MeV. (There are other reasons for believing that m is closer to 500 MeV, and a reasonable error estimate is ±200 MeV around this value.)

The way we have organized this paper may lead the reader of Section 2 to believe that the gluon mass m has been introduced as an <u>ad hoc</u> construct. But it must be remembered that in Section 4 we will provide the technical grounds which suggest the necessity for an effective gluon mass.

2. Qualitative Aspects of Confinement

Here we give a series of arguments that:

(a) Quark confinement requires a vacuum full of fluctuating color fields, and these field strengths have short-range correlations only [6,7,3];

(b) Short-range color field strengths necessitate a <u>long-range pure-gauge</u> part to the vector potential;

(c) The long-range pure-gauge part carries magnetic flux quantized according to the center of the gauge group [8,2], and quark confinement (or gluon screening) can be understood as a non-Abelian Bohm-Aharonov effect (or lack thereof, for gluons).

We then further argue that these concepts are unified by a quantum (not classical!) vortex which fills the vacuum with a tangle of colored fields (with $\langle g^2 G \cdot G \rangle > 0$), and which both sustains and is sustained by the gluon mass m. These vortices (in d = 4) have a central region formed of a closed (possibly at infinity) two-dimensional surface, and a thickness $\simeq m^{-1}$ transverse to the closed surface. Their field strengths fall off exponentially in the transverse directions, and the gauge potential has a pure-gauge long-range part. It is straightforward to show that the vortex vacuum leads to an area law for quark Wilson loops and a perimeter law for gluon Wilson loops.

Consider first the argument [6,7,3] that confinement requires a vacuum full of color fields with short-range correlations, which we will crudely simplify by giving it only for Abelian fields. The Wilson-loop expectation value is transformed by Stokes' theorem and the assumption of Gaussian fluctuations:

$$\begin{aligned} W \equiv \langle \exp i g \oint dx_\mu A^\mu \rangle &= \langle \exp i g \int d\, \Sigma_{\mu\nu} F^{\mu\nu} \rangle \\ &= \exp\left\{ -\frac{1}{2} g^2 \int d\, \Sigma_{\mu\nu}\, d\, \Sigma'_{\alpha\beta} \langle F_{\mu\nu} F'_{\alpha\beta} \rangle \right\} . \end{aligned} \tag{1}$$

Suppose that the field strengths have exponentially-falling correlations, in Euclidean space, which we illustrate with an approximate form of the Green's function:

$$\langle F_{\mu\nu} F'_{\alpha\beta} \rangle \simeq \frac{1}{12} \langle F \cdot F \rangle (\delta_{\mu\alpha}\delta_{\nu\beta} - \delta_{\nu\alpha}\delta_{\mu\beta})\, e^{-m|x-x'|} \tag{2}$$

Then, for a Wilson loop lying in a plane and with dimensions $\gg m^{-1}$, (1) becomes

$$W = \exp\left\{ -\frac{\pi}{2m^2} \langle g^2 F \cdot F \rangle \times \text{area} \right\} \tag{3}$$

This "derivation" of the area law correctly shows the connection of a non-zero (positive!) expectation value $\langle F \cdot F \rangle$ and short-range correlations with confinement, but it has a number of flaws. It is valid only for Abelian theories, and can be misinterpreted to suggest confinement of <u>all</u> group representations in a non-Abelian gauge theory, when in fact representations where the center of the group is trivially represented show screening (perimeter law) and not confinement.

Indeed, a naive argument based on similar lines shows that all representations ought to exhibit a perimeter law: instead of using Stokes' theorem, use the original form of W and replace (2) by $\langle A_\mu A_\nu{}' \rangle \sim \delta_{\mu\nu}\, e^{-m|x-x'|}$. But this is incorrect, since short-range field strengths imply long-range parts to A_μ. To see this, suppose a given component of $F_{\mu\nu}$, say F_{xy}, is equal to $e^{-m\rho}$ with $\rho^2 = x^2+y^2$. Then the total flux Φ through the x-y plane is not zero, and it can be expressed, by Stokes' theorem, as a line integral at infinity:

$$\Phi = \oint d\underset{\sim}{s} \cdot \underset{\sim}{A} \tag{4}$$

Clearly, $\underset{\sim}{A}$ has a <u>long-range</u> part (as well as a short-range part) of the form

$$\underset{\sim}{A} = (\Phi/2\pi)\, \underset{\sim}{\nabla}\phi + \ldots \tag{5}$$

where ϕ is the azimuthal angle in the x-y plane.

We see here a generalization of the Higgs-Goldstone machinery. No matter how gauge fields become massive, whether by symmetry breakdown or otherwise, there must be accompanying long-range pure-gauge excitations (Goldstone modes, when symmetry breaking does occur). These excitations, as Schwinger pointed out long ago, may be dynamical in nature and not explicitly expressed in the Lagrangian, and that is just what happens in QCD.

Let us see how the pure-gauge parts are associated with vortices when the gluon has mass, and how a condensate of these vortices leads to confinement. We will use [2,9] an <u>effective</u> Lagrangian for massive gluons which is locally gauge-invariant and which summarizes the most important part of the quantum-mechanical corrections to the QCD Lagrangian (which is written in Euclidean space):

$$L = \frac{1}{2}\,\mathrm{Tr}\, G_{\mu\nu}{}^2 + m^2\, \mathrm{Tr}[A_\mu - g^{-1} U \partial_\mu U^{-1}]^2 \tag{6}$$

where

$$A_\mu = \frac{1}{2} \Sigma \lambda_a A_\mu^{\ a} , \quad \mathrm{Tr} \frac{1}{2} \lambda_a \frac{1}{2} \lambda_b = \frac{1}{2} \delta_{ab} \tag{7}$$

$$U(\theta) = \exp \frac{i}{2} \Sigma \lambda_a \theta^a \tag{8}$$

This Lagrangian is invariant under local gauge transformations:

$$A'_\mu = V A_\mu V^{-1} - g^{-1} (\partial_\mu V) V^{-1} \tag{9}$$

$$U' \equiv U(\theta') = VU \tag{10}$$

The needed pure-gauge parts are in the matrices U, whose equations of motion are

$$D_\mu (A^\mu - g^{-1} U \partial^\mu U^{-1}) = 0 \tag{11}$$

That they are long-range follows from the perturbative solution [9]

$$\theta_a = g \, \Box^{-1} \, \partial_\mu A_a^{\ \mu} + \ldots \tag{12}$$

There are also non-perturbative solutions to the equations; these are vortices.

A vortex solution to these equations is [2]:

$$A_o = 0 \; ; \quad A_i = g^{-1} Q \hat{\phi}_i \{\rho^{-1} - mK_1(m\rho)\} \tag{13}$$

$$\frac{1}{2} \Sigma \lambda_a \theta_a = Q\phi \tag{14}$$

Here Q is a generator of the gauge group, with $\exp(2\pi iQ)$ in the center of the group (so that a test gluon carried around the vortex at large distances has single-valued field strength), ϕ is the azimuthal angle around the z-axis, and K_1 is a Hankel function of imaginary argument. The long-range gauge part is the first term in curly brackets in (13); the second term decreases exponentially. Although the pure-gauge part is singular by itself at $\rho = 0$, the singularity in A_μ is exactly cancelled by the $mK_1(m\rho)$ term. This has the consequence that $G_{\mu\nu}$ (which decreases exponentially at large ρ) receives no contributions from the singularity in the pure-gauge part. The action per unit extension of the vortex, given by the integral of (6) over the x-y plane, receives a finite contribution from the $G_{\mu\nu}^2$ term, and a logarithmically-singular contribution of the form

$$\int_0 \frac{d\rho}{\rho} \frac{m^2}{g^2} \tag{15}$$

from the mass term. This singularity is an artifact of the effective Lagrangian, and when quantum corrections are included (see Section 4) m and g in (15) must be replaced by running values $m(\rho)$, $\bar{g}(\rho)$. It is required by the dynamical nature of mass generation that $m(\rho)$ vanishes for small ρ, and it then turns out that the quantum action per unit extension is finite.

Now let us see how such a vortex, or rather a condensate of them, leads to a picture of confinement and screening which is even quantitative (up to errors of ±30% or so). First, note that it is easy to generalize the results in (13) and (14) to describe a vortex with any sufficiently large smooth closed two-surface as its core; explicit formulas are given in [2]. (The core of a vortex is the surface on which the long-range pure-gauge part is singular, that is, the z-t plane in (13)-(14).) A large surface of fixed area has a configurational entropy proportional to its extension (i.e., area) and the question of whether a vortex condensate forms or not is that of Kosterlitz and Thouless: is the entropy/extension greater than the action/extension? If so, a condensate forms and QCD is in the confining phase, with $\langle G \cdot G \rangle > 0$; if not, QCD is in a Higgs phase. As we have said, the action is finite and calculable, but it is somewhat trickier to find the entropy because one really needs the entropy at fixed action. Very sharp bends in the core raise the action considerably, so a vortex can only be twisted so far. Estimates (which will not be given here) suggest that the entropy does exceed the action, so a condensate is most likely predicted by this approach. Moreover, a finite fraction of these vortices will have cores of indefinitely large extension, because the greater the extension the greater the difference between entropy and action.

also unity for gluons ($2\pi/3 \to 2\pi$), so in this approximation gluons do not see vortices. We will rectify this in Section 3.

Doing the sums yields

$$W = \exp\left\{ [2 \cos(2\pi/3) - 2]\, U^{-1} e^{-\tilde{A}} \int d^2a \right\} \tag{18}$$

and the quark string tension is

$$K_F = 3U^{-1} e^{-\tilde{A}} \tag{19}$$

In [3], crude calculations of $U^{-1}e^{-\tilde{A}}$ led to $K_F \simeq (\pi/9)\, m^2$ which for $m = 500$ MeV is about 1/2 the accepted value. The idea is simply that $U^{-1}e^{-\tilde{A}}$ is the areal density of vortices, or a bit less than $m^2/2\pi$.

Evidently the vortex condensate encompasses the points (a), (b), (c) made at the beginning of this Section: they furnish an explicit realization of a vacuum filled with a tangle of color fields of short-range correlation. One can, for example, compare (19) with the non-Abelian version of (3) in which the exponent is divided by $2N = 6$ (N is the number of colors and 2N is the ratio of $\mathrm{Tr}\left(\frac{1}{2}\lambda_a \frac{1}{2}\lambda_b \right)$ to Tr I) to estimate, with $U^{-1}e^{-\tilde{A}} \simeq m^2/2\pi$

$$\langle g^2 G\cdot G\rangle \simeq \frac{108}{\pi^2}\, m^4 \simeq 0.7 \text{ GeV}^4 \tag{20}$$

for $m = 500$ MeV. But the vortex condensate goes beyond these points in predicting screening, or a perimeter law, for gluons. We take up this matter of screening in the next Section, and then conclude with more technical arguments which self-consistently relate the gluon mass and the vortex condensate.

3. The Breakable String Potential Between Gluons

Quarks are confined because their fractional color charge cannot be screened by vacuum fluctuations. Gluons can be screened, however, and the gluon-gluon potential stops rising when enough energy has been stored in the vacuum to materialize a new gluon pair, the elements of which bind to the original gluons to form glueballs. The necessary

Consider a large quark Wilson loop which is topologically linked to a vortex core. (This is easy to visualize in three dimensions, where a vortex core is a closed loop like a Wilson loop.) If the loop winds once around the core, it is easy to see that

$$\mathrm{Tr}\ P \exp i\ g \oint dx_\mu A^\mu = \mathrm{Tr}\exp 2\pi i Q = z\ \mathrm{Tr}\ I \tag{16}$$

where the numbers z give the one-dimensional representation of the center of the gauge group ($z = e^{\pm 2\pi i/3}$ for SU(3)). For N vortices linked once (or one vortex linked N times, etc.), $z \to z_1 z_2 \ldots z_N$. In (16), we ignore terms vanishing exponentially with the minimum distance from core to Wilson loop, but those contain the perimeter law for gluons, discussed in Section 3. For an unlinked vortex, or a gluon Wilson loop, z is replaced by unity. It is possible, with the aid of the general formulas for a vortex potential [2], to express (16) in terms of the topological integrals which give the linking number of the loop and vortex core [10].

The essential properties of quark confinement and gluon screening now follow readily, just as they do in Abelian models [11,12], in the dilute-vortex approximation. For a planar quark Wilson loop, its expectation value can be roughly expressed in two-dimensional form, again ignoring possible contributions from perimeter terms:

$$\begin{aligned} W &= \frac{1}{3} \langle \mathrm{Tr}\ P \exp i\ g \oint dx_\mu A_\mu \rangle \\ &= Q^{-1}\ \Sigma (N_+!)^{-1}\ (N_-!)^{-1} \int \frac{d^2 a_1}{U} \ldots \exp\left\{ \frac{2\pi i}{3}\ (N_+ - N_-) - \tilde{A}(N_+ + N_-) \right\} \end{aligned} \tag{17}$$

where Q is the partition function, $N_\pm$ is the number of vortices with $z = \exp \pm 2\pi i/3$ linked to the loop, and $\tilde{A}$ is a combination of the vortex action and configurational entropy. The remaining entropy is expressed as the integral $\Pi(U^{-1} \int d^2 a_i)$ over the position a_i where the ith vortex intersects the Wilson loop; U is a normalization area which arises in a very difficult collective-coordinate determinant. The factors $\exp \pm(2\pi i/3)\ N_\pm$ become unity for an unlinked vortex, just as they do for the vacuum, so the integral $\int d^2 a$ extends only over the interior of the Wilson loop because outside vortices are considered to be unlinked and their contribution is cancelled by Q. These factors are energy is of the order of 2m. This process has been observed on the lattice by Bernard [13], who finds that the gluonic potential becomes flat at about 1.1 GeV. At lower energies (smaller distances) the string potential rises roughly linearly with a slope K_A that is roughly $2K_F$ (or possibly $2N^2(N^2-1)^{-1} K_F$). The factor of two can be understood by thinking of the gluon as a quark-antiquark pair, so gluons are joined by two quark strings which can reconnect.

Let us study this process in the dilute-vortex-gas approximation, as we did for quark confinement. Just as for that case there will be many uncertainties connected with the approximation, but we hope that most of them can be reduced by normalizing our calculations to K_F.

Consider a rectangular Wilson loop of sides T, R with $T \gg R, m^{-1}$. We define

$$W = \frac{1}{8} \langle \mathrm{Tr}\, P\, e^{ig \oint dx_\mu A_\mu} \rangle = \exp\{-TV(R)\} \tag{21}$$

where the trace is now in the adjoint representation of SU(3), and V(R) is the gluon-gluon potential arising from vortices. At large R, $V(R) \rightarrow$ const. so W shows a perimeter law. The calculation is made strictly two-dimensional by using the form (13)-(14) for the vortex potential, where $\hat{\phi}_i$ is in the R-T plane. In the adjoint representation, the matrix Q has eigenvalues (1,1,-1,-1,0,0,0,0), and it is evident that the $Q\phi_i(g\rho)^{-1}$ term in the vortex potential contributes nothing to $\mathrm{Tr}\, P \exp i \oint dx^\mu A_\mu$, whether the vortex is linked or not. In evaluating the contribution of the short-range part of the vortex to this integral, we need

$$\int_{-\infty}^{\infty} \frac{dy\, mx}{(x^2+y^2)^{1/2}} K_1(m(x^2+y^2)^{1/2}) = \pi e^{-mx} \tag{22}$$

where the integral along y represents the contribution from one long side of the Wilson loop in the limit $T \rightarrow \infty$, and x is the distance of the vortex core from the long side. This shows the short-range nature of the vortex-gluon interaction. After some manipulations similar to those which yielded (17)-(19) for quarks, we find (details will be reported elsewhere)

$$V(R) = \frac{2}{3} K_F m^{-1} \int_0^{\infty} dy\{1 - \cos[\pi e^{-y} - \pi e^{-\xi-y}]\}$$
$$+ \frac{1}{3} K_F m^{-1} \int_0^{\xi} dy\{1 - \cos[\pi e^{-y} + \pi e^{-\xi+y}]\} \quad (23)$$

where $\xi = mR$, and (19) has been used to normalize the calculation. The first term of (23) comes from vortices which are topologically unlinked, the second term from the linked vortices.

Figure 1 shows a plot of V(R) out to $\xi \simeq 3$, where V reaches its maximum value of 2.4 $K_F m^{-1}$. This should be of the order of 2m, that is, K_F should be about 0.8 m^2 which puts m in the range 400-500 MeV. The average linear rise of V(R) (which vanishes like R^2 for very small R) corresponds to $K_A \simeq 1.5\ K_F$, about 30% lower than either the naive guess of $2K_F$ made earlier or lattice calculations [13].

A potential of this general type has been used [5] to calculate the masses of some of the low-lying glueballs; details are reported by A. Soni in this volume.

4. Dynamical Mass From Schwinger-Dyson Equations

We want to see how the strong interactions of QCD generate a dynamical gluon mass--a mass that vanishes at large momentum or small distances. Just as in other contexts of dynamical mass generation [4,9], this behavior of the mass reflects the attractiveness of forces in the channel carrying the quantum numbers of the mass, that is, color-singlet 0^{++}, and mass generation in QCD is closely tied to the strong binding of the 0^{++} glueball.

It is tempting to study the dynamics of the gluon propagator [14], as Ball reports in this Conference, but unfortunately the propagator is gauge-dependent. Instead, we study a related Green's function which is gauge-invariant and which represents a partial resummation of the Feynman graphs for another gauge-invariant Green's function. This new "propagator" is best thought of as expressing part of the force law between color-octet test particles.

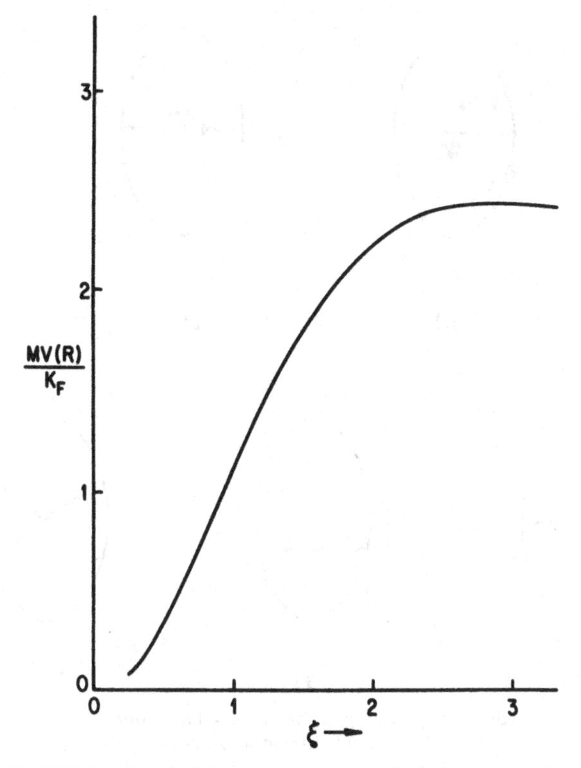

FIGURE 1 Breakable string potential between gluons.

Let us invent an octet of very heavy test fermions ψ, and write the gauge-invariant Green's function

$$G(x,y) = \langle T(\overline{\psi}_a(x)\ \psi_a(x)\ \overline{\psi}_b(y)\ \psi_b(g))\rangle \tag{24}$$

(Note: In [3] the same Green's function was considered, but erroneously labeled as referring to spin-zero test particles. For spin-zero particles the same final results emerge, but seagull graphs must be accounted for.) Normally G, which is gauge-invariant, is decomposed into gauge-variant Feynman graphs of which are shown in Figure 2, but it would be better if it could be written as a sum of gauge-invariant pieces. It turns out that we can go as far in this direction for QCD as we can for QED, writing a new propagator whose only gauge dependence is in its free part; the vacuum polarization tensor is fully gauge-invariant.

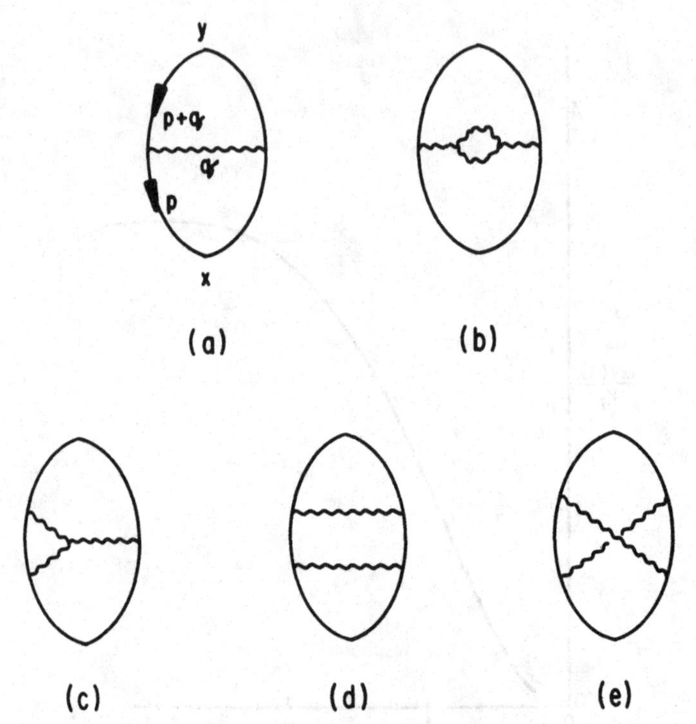

FIGURE 2 Some Feynman graphs which have one-particle parts in the light-cone gauge.

Although any gauge may be used with the same final outcome it is very convenient to use the light-cone gauge $n\cdot A = 0$, $n^2 = 0$ (so we must begin in Minkowski space). This is because there are no ghosts, hence naive Ward identities hold, and because momentum-space integration is as simple as it is for covariant gauges [9]. The latter is not true for axial gauge with $n^2 \neq 0$. In this gauge the free propagator is

$$\Delta_{\mu\nu} = \frac{P_{\mu\nu}}{q^2+i\varepsilon}\ , \quad P_{\mu\nu} = -g_{\mu\nu} + \frac{n_\mu q_\nu + n_\nu q_\mu}{n\cdot q} \tag{25}$$

The longitudinal parts of (25) play a special role: when q_μ fits a test-particle vertex it produces terms which cancel out certain test-particle propagators via the elementary Ward identity

$$q_\mu \gamma^\mu = S^{-1}(p+q) - S^{-1}(p) \tag{26}$$

where S is the test-particle propagator. The three-gluon vertex also has longitudinal parts which do the same thing. As a result, the vertex corrections and ladder graphs of Figure 2 have parts which are attached to each of the test-particle lines at only a single point, just like a propagator; these parts of graphs are shown in Figure 3. The sum of all these one-particle parts, which we call $\hat{\Delta}_{\mu\nu}$, is necessarily gauge-independent (or n_μ-independent, in the light-cone gauge), and explicit one-loop calculations [3] show this to be the case.

This much was known some years ago [15]. The new step is to recognize that the new "propagator" $\hat{\Delta}_{\mu\nu}$, defined as the sum of all orders of the one-particle parts, obeys a Schwinger-Dyson equation. This equation requires a new vertex $\hat{\Gamma}_{\mu\nu\alpha}$, constructed much as $\hat{\Delta}_{\mu\nu}$ is, and whose Ward identity involves the difference of two $\hat{\Delta}^{-1}_{\mu\nu}$ (cf. (26)). In principle, one should write down another equation for the new vertex, and ultimately end up with an infinite set of coupled equations for an array of modified Green's functions.

In practice, this is impossible and we resort to the so-called gauge technique, which expresses the longitudinal part of the vertex directly in terms of the propagator. This was investigated for QED [15,16], where it turns out to linearize the electron propagator equation. Such is not the case of QCD [3,14], but the technique is still very useful. The reasons are that the neglected transverse part of the vertex vanishes (for a theory with a mass gap) at least one

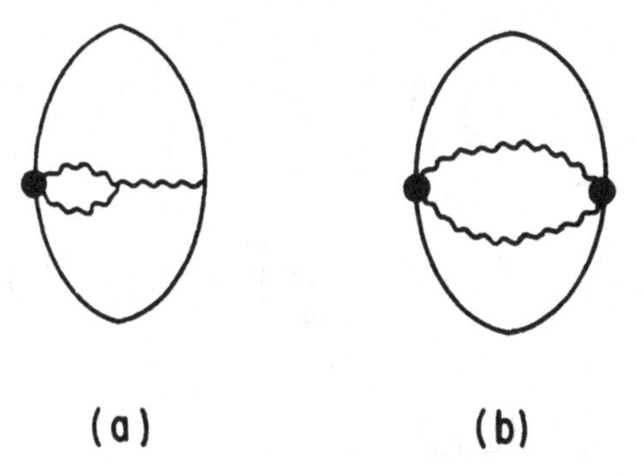

FIGURE 3 One-particle parts of the two-gluon graphs of Figure 2.

power of q faster than the longitudinal part in the infrared regime $q \simeq 0$, and the transverse part has one more power of g^2 than the longitudinal part, so it should be negligible in the ultraviolet. Actually, the latter is not really true: neglecting the transverse part in the ultraviolet, which is formally an error of $O(g^4)$, does not lead to the correct renormalization-group-improved perturbative results at large q. In practice this neglect forces one to work with erroneous values of the lowest-order renormalization-group coefficients, e.g., in $\beta = -bg^3 + \ldots$, one finds b multiplied by factors like 16/11 [14] or 2 [3] compared to the correct value.

Since the dynamical equation for $\hat{\Delta}_{\mu\nu}$ is independent of n_α, we can write it in Euclidean space. Define the Minkowski-space version of $\hat{\Delta}_{\mu\nu}$ through a Källen-Lehmann representation:

$$\hat{\Delta}_{\mu\nu} = P_{\mu\nu} \int \frac{d\lambda^2 \rho(\lambda^2)}{q^2 - \lambda^2 + i\varepsilon} \tag{27}$$

and the Euclidean version of the scalar part by

$$d(q^2) = \int \frac{d\lambda^2 \rho(\lambda^2)}{q^2 + \lambda^2} \qquad (q^2 > 0) \tag{28}$$

We need to eliminate the modified vertex $\hat{\Gamma}_{\mu\nu\alpha}$ in order to arrive at a self-contained non-linear equation for d. The gauge technique expresses $\hat{\Gamma}_{\mu\nu\alpha}$ in terms of an integral over $\rho(\lambda^2)$, including long-range pure-gauge parts which we know must be present if mass is to be generated. After lengthy calculations, the equation for d is

$$\begin{aligned} d^{-1}(q^2) = q^2 \left[K + bg^2 \int_0^{q^2/4} dz \left[1 - \frac{4z}{q^2} \right]^{1/2} d(z) \right] \\ + \frac{bg^2}{11} \int_0^{q^2/4} dz\, z \left[1 - \frac{4z}{q^2} \right]^{1/2} d(z) + d^{-1}(0) \end{aligned} \tag{29}$$

$$d^{-1}(0) = \frac{3bg^2}{11\pi^2} \int d^4k\, d(k^2) \tag{30}$$

The number K depends on the renormalization point μ; it is chosen so $d^{-1}(\mu^2) = \mu^2$.

We now come to an extremely important point. Equation (30) is a gauge-invariant way of expressing the gluon mass through the seagull graph (which by itself is not gauge-invariant, but is modified to a gauge-invariant form by the time one has gotten to (29) and (30)). Evidently the right-hand side of (30) is proportional to some gauge-invariant version of $\langle\Sigma[A_\mu^a(x)]^2\rangle$, and equally evidently some such vacuum expectation value must exist if $\langle G\cdot G\rangle \neq 0$. Of course, any of these expressions needs regulation, and we will not discuss here in detail the various efforts [3] in this direction. It turns out that saturating either of the above expectation values with a dilute vortex gas yields finite, calculable values for them, and it is in this sense that a vortex condensate is self-consistently related to the gluon mass (see also [17]). That is, the gluon mass generates the vortices as in (13)-(14), while the vortex gas generates the mass via (30).

These considerations do not yet establish the existence of a non-zero mass; perhaps (29) and (30) have massless solutions ($d^{-1}(0) = 0$) for which $\langle G\cdot G\rangle = 0$. Such solutions, if they exist, would be consistent with a dimensionally-regularized value of zero for the right-hand side of (30). In fact, a detailed study of (29), with $d^{-1}(0)$ taken as arbitrary (including zero), shows that no solutions exist for $d^{-1}(0)$ sufficiently small (e.g., much less than Λ_{RG}^2). This is because the famous Landau ghost of massless perturbation theory begins to emerge and cannot be made consistent with (29). Within the approximations that led to (29), there must be a gluon mass (equivalently, $d^{-1}(0) = 0$). This is one of the main points of this Section.

The other main point is that one can predict from (29) how the mass term behaves at large q^2. One finds

$$m(q^2) \sim (\ell n\, q^2)^{-12/11} \tag{31}$$

which is decreasing fast enough to give vortices finite action (see the remarks below (15)). As we have already said, this decrease is demanded by the existence of attractive forces in the 0^{++} glueball channel, which suggests that an alternate way of understanding dynamical mass generation is through this glueball and its vacuum expectation value. This is indeed the case [3], but we have no room for details here.

We close by remarking that the ultraviolet difficulties arising from neglect of the transverse vertex can be overcome. King [18] has shown in detail how this works in QED by explicitly constructing a transverse vertex which correctly yields renormalization-group-improved perturbation theory in the ultraviolet, which vanishes like a power of q in the infrared, and which leaves the electron propagation equation linear. King and the author are now working on analogous results for the $\hat{\Delta}_{\mu\nu}$ equations.

5. Acknowledgements

I am happy to thank Professor San-Fu Tuan for his hospitality at the University of Hawaii during the writing of this manuscript. The work itself was supported by the National Science Foundation.

References

[1] J. M. Cornwall, in Deeper Pathways in High-Energy Physics, ed. B. Kursonoglu, A. Perlmutter, and L. Scott (Plenum, New York, 1977), p. 683.

[2] J. M. Cornwall, Nuc. Phys. B157, 392 (1979).

[3] J. M. Cornwall, Phys. Rev. D26, 1453 (1982).

[4] E. Farhi and R. Jackiw, editors, Dynamical Gauge Symmetry Breaking (World Scientific, Singapore, 1982).

[5] J. M. Cornwall and A. Soni, Phys. Lett. 120B, 431 (1983).

[6] H. B. Nielsen and P. Olesen, Niels Bohr Institute Report NBI-HE-79-45, 1979 (unpublished).

[7] R. P. Feynman, talk at the Lisbon Conference, July 1981.

[8] G. 't Hooft, Nuc. Phys. B138, 1 (1978).

[9] J. M. Cornwall, Phys. Rev. D10, 500 (1974).

[10] H. Flanders, Differential Forms (Academic Press, New York, 1963).

[11] C. G. Callan, Jr., R. Dashen, and D. Gross, Phys. Lett. 66B, 375 (1977).

[12] T. Banks, R. Myerson, and J. Kogut, Nuc. Phys. B129, 493 (1977).

[13] C. Bernard, Phys. Lett. 108B, 431 (1982).

[14] R. Anishetty, M. Baker, S. K. Kim, J. S. Ball, and F. Zachariasen, Phys. Lett. 86B, 52 (1979).

[15] J. M. Cornwall and G. Tiktopoulos, Phys. Rev. D15, 7937 (1977).

[16] R. Delbourgo and P. West, J. Phys. A10, 1049 (1977).

[17] R. J. Kares and M. Bander, Phys. Lett. 96B, 320 (1980).
[18] J. E. King, Phys. Rev. D, 1983 (to be published).

ANALYTIC REGULARIZATION AND RENORMALIZATION OF NONPERTURBATION THEORIES

H.C. Lee* and M.S. Milgram*

1. Introduction

Quantum field theories suffer from infinities. In perturbation theories, these infinities manifest themselves as ultraviolet (UV) divergences in Feynman integrals. In massless theories, there are also infrared (IR) divergences to contend with. In perturbation expansion, the order-by-order removal of these infinities - the renormalization program[1] - is well understood. The program has been tremendously simplified since the advent of dimensional regularization[2,3], a technique whereby the infinities are analytically isolated as poles in the complex ω-plane, where 2ω is the generalized dimension of Euclidean space-time. For nonperturbation theories, a general and viable renormalization procedure has not yet been devised. The problem with which we shall be concerned here is the regularization and renormalization of a nonperturbation theory as represented by the nonlinear equations derived from it. We will describe a technique that should allow one to _analytically_ regulate and ultimately renormalize the equation.

2. Loop expansion, overlapping divergences and polylogs

To understand the nature of the problem consider a perturbation expansion ordered by the number of loops in the Feynman diagrams. Corresponding to each loop is an integral possibly containing a divergence with which a logarithm of an external momentum p, $\ell n p^2$, is associated. In dimensional regularization, the symbiotic relation between the infinity and the logarithm is embodied in the expression $(p^2)^{\varepsilon}/\varepsilon$, where $\varepsilon \equiv \omega-2$. For example, a two-point function may have poles of $O(1/\varepsilon)$ and a $\ell n p^2$ dependence at the one-loop level (Fig. 1a), poles of $O(1/\varepsilon^2)$ and a $\ell n^2 p^2$ dependence at the two-loop level (Fig. 1b), and so on. The poles can be absorbed into renormalization constants and eventually be cancelled by counterterms, whereas the logarithms determine the anomalous p-dependence of the two-point function. A difficulty arises when, in addition to poles and logarithms, an N-loop (Fig. 1c, with $N > 2$)

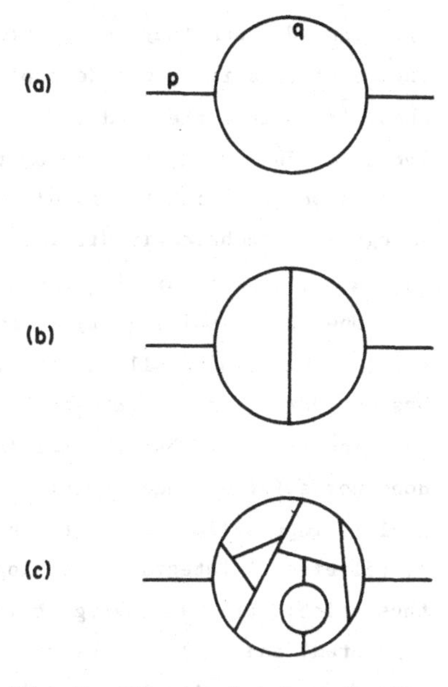

Fig. 1. Divergent integrals for N-loop diagrams generate logarithms of order N.

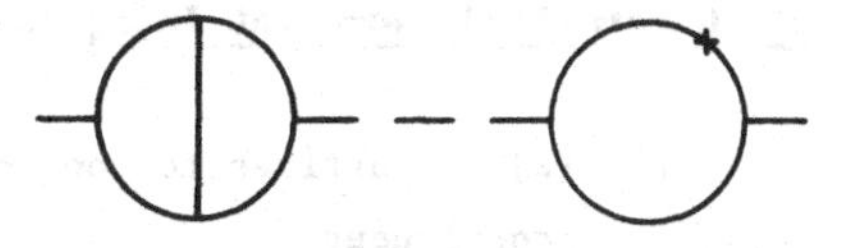

Fig. 2. Two-loop diagram and one-loop subtraction inserted with counterterm (×).

integral has overlapping divergences - terms of $O(\ln^m p^2/\varepsilon^{N-m})$, $1 \le m \le N-1$ - that are infinite but unrenormalizable. Such terms are removed with a prescription due to 'tHooft and Veltman[2]: corresponding to any N-loop integral, subtract (N-1)-loop integrals inserted with the appropriate counterterms needed for the theory to be finite at the (N-1)-loop level. For example, a two-loop integral with its subtraction (Fig. 2) appear schematically as

$$\int d^{2\omega}q \int d^{2\omega}q'(\cdots) - \frac{1}{\varepsilon}\int d^{2\omega}q(\cdots)$$

$$= \int d^{2\omega}(\cdots)\left[\frac{(q^2)^\varepsilon}{\varepsilon} - \frac{1}{\varepsilon}\right], \qquad (1)$$

where divergent integrals are implicit, and the first term in the square bracket results from the integration over q'. In summary, after regularization and removal of pole terms with logarithmic residues according to the 'tHooft-Veltman prescription, the (renormalized) two-point function becomes a power series in p^2 and $\ln p^2$ - a "polylog" in p^2. Generally, physical quantities in renormalizable perturbation theories are polylogs of the momenta, masses or other dimensional quantities of the fields involved.

3. Schwinger-Dyson Equation for Gluon Propagator

A nonlinear equation of considerable current interest is the simplified Schwinger-Dyson equation for the gluon[4,5],

$$Z^{-1}(p,n) = 1 + g_0^2 \int d^4q\, K(p,q,n)\, Z(q,n) + \cdots \qquad (2)$$

where Z is the reduced gluon propagator, n is an auxiliary vector defining the axial gauge[6] condition $A\cdot n = 0$ and K is a known kernel. As far as regularization is concerned the terms represented by ··· in (2) are ignored since they do not present any additional difficulty. The choice of an axial gauge decouples the gluon field from Faddeev-Popov ghosts[7] so that the need to solve coupled nonlinear equations is avoided. In return, because of the presence of axial gauge singularities associated with the condition $q\cdot n = 0$, the regularization of the integral is technically difficult.[8] Generally, the integral in (2) may be UV, IR and "axial gauge" (AX) singular.

One way of solving (2) is to assume Z to be some series in p^2 and $p\cdot n$ (our discussion will ignore the factor n^2 for, being a constant, it has no bearing on the integral). Our experience with perturbation theories suggests that the series be a polylog; a simple power series does not suffice since logarithms are generated when the divergent integral is regularized. In turn such an expansion calls for the evaluation of (divergent) integrals with logarithmic integrands. Two questions thus arise: (i) Can divergent integrals with logarithmic integrands be regulated? and (ii) Are the regulated integrals renormalizable - i.e. free from poles with logarithmic residues? The answers, "yes" and "yes", are explained in the following sections.

4. Generalized Two-point Integrals in an Axial Gauge

For (2) it suffices to consider the class of integrals (s=0 or 1, ω, κ, μ, ν continuous)

$$S_{2\omega}(p,n;\kappa,\mu,\nu,s) = \int d^{2\omega}q\,[(p-q)^2]^{\kappa}(q^2)^{\mu}(q\cdot n)^{2\nu+s} \quad . \tag{3}$$

Within the region in the (ω,κ,μ,ν)-space delineated by

$$\omega+\kappa+\mu+\nu < 0, \quad \omega+\kappa > 0, \quad \omega+\mu+\nu+s > 0, \quad 2\nu+s \geq -1, \tag{4}$$

(and including the point $\omega=2$, $\kappa = \mu = \nu = -s = -1$) the integral is regular and is given by the representation[9]

$$S_{2\omega}(p,n;\kappa,\mu,\nu,s) = \frac{\pi^{\omega}(p^2)^{\alpha_1}(n^2)^{\alpha_2}(p\cdot n)^{s}\,\Gamma(\alpha_2+s+1/2)}{\Gamma(\beta_1-\alpha_0)\Gamma(\beta_1-\alpha_1)\Gamma(-\alpha_0-\alpha_1-s)\Gamma(-\alpha_2)}$$

$$\times\, G^{2,3}_{3,3}\left(y\,\middle|\,\begin{matrix}1+\alpha_0,1+\alpha_1,1+\alpha_2;\\ 0,\ \beta_1;\ 1/2-s\end{matrix}\right), \tag{5}$$

where $y = (p\cdot n)^2/(p^2n^2)$, $\alpha_0 = -(\omega+\mu+\nu+s)$, $\alpha_1 = \omega+\kappa+\mu+\nu$, $\alpha_2 = \nu$, $\beta_1 = \omega+\kappa+\nu$ and G is a Meijer G-function[10] which is a well-defined function of all values of its arguments. By the principle of analytic continuation the right-hand side of (5) can be used to represent the integral in (3) over the whole (ω,κ,μ,ν)-space. The derivation of (5), based on a hybrid of dimensional and analytic regularization techniques, is non-trivial and has been described in detail elsewhere.[9] Here we only point out that the AX singularity is analytically regulated by Euler's representation for the gamma function

$$[(q\cdot n)^2]^\alpha = \frac{1}{\Gamma(-\alpha)} \int_0^\infty t^{-\alpha-1} e^{-(q\cdot n)^2 t} dt \tag{6}$$

rather than by the commonly and widely used principal-value prescription[11] (which leads to notoriously difficult computations).

"Primal" integrals in four-dimensional space-time, corresponding to $\omega=2$ and (κ,μ,ν) = integers in (3), are obtained from (5) by letting $(\kappa,\mu,\nu) \to$ integers and $\varepsilon \equiv \omega-2$ finite, and then allowing ε to approach zero.

5. Exponent derivatives

Primal integrals with logarithmic integrands, discussed earlier, are evaluated by taking "exponent derivatives" of the analytic representation (5):

$$\int d^{2\omega}q\,[(p-q)^2]^K (q^2)^M (q\cdot n)^{2N+s} \ln^{K'}(p-q)^2 \ln^{M'} q^2 \ln^{N'}(q\cdot n)^2$$

$$= \lim_{\kappa\to K,\,\mu\to M,\,\nu\to N} \left(\frac{d}{d\kappa}\right)^{K'} \left(\frac{d}{d\mu}\right)^{M'} \left(\frac{d}{d\nu}\right)^{N'} S_{2\omega}(p,n;\kappa,\mu,\nu,s). \tag{7}$$

The analogy between an exponent derivative and the 'tHooft-Veltman prescription for treating overlapping divergences is exposed by comparing the square bracket in (1) to the relation

$$\ln p^2 = \lim_{\mu\to 0} \frac{1}{\mu}\,[(p^2)^\mu - 1]. \tag{8}$$

6. Poles and Renormalizability

The representation (5) is a meromorphic function of the indices α_0, α_1, α_2 and β_1 such that the primal integrals have poles of $O(1/\varepsilon)$ necessarily when at least one of 2+K+M+N, -(2+M+N+s) and -(2+K) is a non-negative integer. These three cases correspond respectively to the original Feynman integrals being UV divergent, or IR divergent at either q=o or q=p. Significantly, the representation is free from AX singularities at ν = negative integers. Furthermore, it can be shown[9] that a prescription for exponent derivatives of the form (7) exists which gives rise to only those high-order poles and associated logarithms generated from the relation

$$\left(\frac{d}{d\varepsilon}\right)^r \left[\frac{(p^2)^\varepsilon}{\varepsilon}\right] = \frac{(-)^r r!}{\varepsilon^{r+1}} + \frac{1}{r+1} \ln^{r+1}(p^2) + O(\varepsilon). \tag{9}$$

In other words, if the two-point function Z, as a solution of (2), gives rise to regulated primal integrals (5) and exponent derivatives (7), then the infinite part of Z is composed of poles which, having only non-logarithmic residues, can be removed by counterterms. Z is therefore renormalizable analytically.

7. An Example

General results covering all primal integrals enveloped by (5) are given elsewhere.[9] Here, as an example, we evaluate the integral

$$I = \int d^4q \frac{[p^2q^2 - (p\cdot q)^2]}{(p-q)^2 q^2 (q\cdot n)^2 [(p-q)\cdot n]^2} \tag{10}$$

encountered[12] in the solution of (2) at the "one-loop" level. Employing partial fractions,

$$\frac{1}{(q\cdot n)[(p-q)\cdot n]} = \frac{1}{p\cdot n}\left[\frac{1}{q\cdot n} + \frac{1}{(p-q)\cdot n}\right], \tag{11}$$

and changing variable,

$$q \rightarrow p-q, \tag{12}$$

when necessary, the integral can be reduced to a sum of primal integrals:

$$I = (p\cdot n)^{-2}\, F(p^2,0) + 2(p\cdot n)^{-3} F(p^2,1) \;, \tag{13}$$

where

$$F(p^2,s) = S(0,0,-1,s) - \frac{1}{2}\, S(1,-1,-1,s) - \frac{1}{2}\, S(-1,1,-1,s)$$

$$+ p^2 S(-1,0,-1,s) + p^2 S(0,-1,-1,s) - \frac{p^4}{2}\, S(-1,-1,-1,s) \;, \tag{14}$$

with the short-hand notation $S(\kappa,\mu,\nu,s) \equiv S_4(p,n;\kappa,\mu,\nu,s)$. Generalizing to 2ω dimensions and using (5), one finds that of the twelve integrals, the six with $\kappa=0$ or 1 vanish identically[9], and the remaining six are:

$$S(-1,1,-1,0) = \frac{2\pi^\omega p^2}{n^2}\left[(1-4y)\left(\frac{1}{\varepsilon} + \ln p^2 + \ln 4y + \gamma\right) + 4y\right], \tag{15a}$$

$$S(-1,0,-1,0) = \frac{2\pi^\omega}{n^2}\left(\frac{1}{\varepsilon} + \ln p^2 + \gamma + \ln 4y\right), \tag{15b}$$

$$S(-1,-1,-1,0) = -\frac{2\pi^\omega}{p^2 n^2}\left(\frac{1}{\varepsilon} + \ln p^2 + \gamma - \ln 4y\right), \tag{15c}$$

$$S(-1,1,-1,1) = -\frac{2\pi^\omega (p\cdot n) p^2}{n^2}(1-2y)\left(\frac{1}{\varepsilon} + \ln p^2 + \gamma + \ln 4y - 2\right) \tag{15d}$$

$$S(-1,0,-1,1) = -\frac{2\pi^\omega (p\cdot n)}{n^2}\left(\frac{1}{\varepsilon} + \ln p^2 + \gamma + \ln 4y - 2\right), \tag{15e}$$

$$S(-1,-1,-1,1) = -\frac{2\pi^2 (p\cdot n)}{p^2 n^2}\left(-2 + \ln 4y + yZ_2/2\right), \tag{15f}$$

where for $|y| \leq 1$

$$Z_m = \sqrt{\pi}\sum_{\ell=0}^{\infty}\frac{y^\ell\,\Gamma(m+\ell)}{\Gamma(m+1/2+\ell)}\left[\psi(m+\ell) - \psi(m+1/2+\ell) + \ln y\right], \tag{16}$$

$y = (p\cdot n)^2/(p^2 n^2)$, $\gamma = 0.577\cdots$ is the Euler-Mascheroni constant and $\psi(x)$ is the digamma function. Of the six integrals, five have infinite parts while the last one (15f) is regular. (Interestingly, of all the primal integrals covered by (3), (15f) is the only one that cannot be expressed as a finite series; it has previously been reduced to a one-fold integral by van Neerven[8] and by Konetschny[8].) However, upon substitution of (15) into (13), all infinite terms cancel, yielding

$$I = \frac{2\pi^2}{n^4}\left(2 + \frac{1}{2} Z_2\right). \tag{17}$$

Note that, from (16), the integral has a $\ln y$ singularity in the limit $y \to 0$. In other words, as far as the integral (10) is concerned, the special gauge[13] $p \cdot n = 0$, p^2 and n^2 finite, is singular.

For exponent derivatives, we give as examples those of $S(-1,-1,-1,0)$:

$$\left.\frac{d}{d\kappa} S(\kappa,-1,-1,0)\right|_{\kappa=-1} = \int d^{2\omega}q \; A \ln(p-q)^2$$

$$= \frac{2\pi^\omega}{p^2 n^2}\left[\frac{1}{\varepsilon^2} + \frac{\gamma}{\varepsilon} + \frac{\gamma^2}{2} - \frac{\pi^2}{4} - \frac{1}{2}\ln^2(p^2/4y) + \ln^2 4y + yZ_1\right], \tag{18a}$$

$$\left.\frac{d}{d\mu} S(-1,\mu,-1,0)\right|_{\mu=-1} = \int d^{2\omega}q \; A \ln q^2$$

$$= \frac{2\pi^\omega}{p^2 n^2}\left[\frac{1}{\varepsilon^2} + \frac{\gamma}{\varepsilon} + \frac{\gamma^2}{2} - \frac{\pi^2}{12} - \frac{1}{2}\ln^2(p^2/4y) + \frac{1}{2}\ln^2 4y + yZ_1\right], \tag{18b}$$

$$\left.\left(\frac{d}{d\nu} - \frac{d}{d\mu} - \ln n^2\right) S(-1,\mu,\nu,0)\right|_{\mu=\nu=-1} = \int d^{2\omega}q \; A \ln\left[(q\cdot n)^2/q^2 n^2\right]$$

$$= \frac{2\pi^\omega}{p^2 n^2}\left\{(\ln 4 - 2)\left[\frac{1}{\varepsilon} + \gamma + \ln(p^2/4y)\right] + \frac{1}{2}\ln^2 4y + yZ_1\right\}, \tag{18c}$$

where $A \equiv \left[(p-q)^2 q^2 (q\cdot n)^2\right]^{-1}$ and Z_1 is given in (16). The particular logarithm in the integrand in (18c) is chosen because the auxiliary vector n appears only in the logarithm $\ln y$. Note that whereas (18a,b) have $1/\varepsilon^2$ and $\ln^2 p^2$ terms, (18c) does not. This is an aspect of the fact that, unlike logarithms of p^2, logarithms of y are artifacts of the axial gauge, and are unrelated to multi-loop effects.

A computer code utilizing the algebraic manipulator SCHOONSCHIP[14] to evaluate primal integrals and their exponent derivatives has been written[15].

8. Multi-point Integrals and Massive Integrals

It is worth pointing out that the nonperturbative solution of the two-point function Z in (2) calls for the evaluation of only two-point integrals (3) - high-order terms being given by exponent derivatives. This contrasts with perturbation theories in which the calculation of a two-point function to high orders involves the evaluation of m-point integrals (integrals with m-1 external momenta), m > 2. There is nevertheless a need to evaluate at least three- and four-point integrals even in nonperturbation theories. The method described briefly in section 4 and in detail elsewhere[9] may be applied to evaluate such integrals. In short, for any massless m-point integral in an axial gauge, the UV divergence can always be isolated as a simple pole and the AX singularity analytically removed, reducing the integral to a nontrivial, (m-1)-fold parameteric integration in which all IR singularities reside. The analytic result for m=2 is given by (5) but, as far as we know, such a result for m > 2 has not been derived.

If all particles are massive then the (m-1)-fold integral mentioned above has no IR divergences. Closed-form expressions for massive integrals are generally more difficult to obtain then those for massless ones; the analytic expression for two-point integrals with one finite mass has been derived,[16] that for two masses has not. Work on these integrals is progressing.

9. Conclusion

A technique for analytically regulating divergent integrals in nonlinear integral equations derived from nonperturbation theories, thereby rendering solutions of the equations analytically renormalizable, is proposed. The technique hinges on one's ability to find analytic representations for generalized Feynman integrals, where the dimension of space-time and the exponents in the integrand are generalized to continuous variables. Such a representation for the class of massless, two-point integrals in ghost-free axial gauges (integrals in covariant gauges form a subclass) has been formed. This representation should in principle permit the Schwinger-Dyson equation for the gluon propagator to be renormalized and solved analytically, at least in the infrared limit. Work in this direction is in progress.

References

* Atomic Energy of Canada Limited, Chalk River Nuclear Laboratories, Chalk River, Ontario, Canada KOJ 1J0

1. For reviews: E.S. Abers and B.W. Lee, Phys. Rep. 9C(1973)1; P. Ramond, "Field Theory" (Benjamin/Cummings, Reading, Massachusetts, 1981).
2. G. 'tHooft and M. Veltman, Nucl. Phys. B44(1972)189.
3. For reviews: G. Leibbrandt, Rad. Mod. Phys. 47(1975)849; S. Narison, Phys. Rep. 84(1982)263.
4. J.S. Ball and F. Zachariasen, Nucl. Phys. B143(1978)148; S. Mandelstam, Phys. Rev. D20(1979)3223; R. Delbourgo, J. Phys. G. 5(1979)603; J.M. Cornwall, Phys. Rev. D26(1982)1453; A.I. Alekseev, Teor. Mat. Fiz. 48(1981)324; M. Baker, J.S. Ball and F. Zachariasen, Nucl. Phys. B186(1981)531,560; W.J. Shoenmaker, Nucl. Phys. B194(1982)535.
5. See also talks by J. Ball and by P. Johnson in these Proceedings.
6. R. Delbourgo, A. Salam and J. Strathdee, Nuovo Cimento 23A(1974)237; W. Konetschny and W. Kummer, Nucl. Phys. B100(1975)106.
7. B.S. DeWitt, Phys. Rev. 162(1967)1195; L.D. Faddeev and V.N. Popov, Phys. Lett. 75B(1967)29.
8. e.g., D.M. Capper and G. Leibbrandt, Phys. Rev. D25(1982)1002,1009; W.L. van Neerven, Z. Phys. C14(1982)241; W. Konetschny, preprint, Tech. U. Wien (March, 1983).
9. H.C. Lee and M.S. Milgram, preprint CRNL-83-III-03 (to be published).
10. Y. Luke, "The Special Functions and their Approximations", (Academic Press, New York, 1969) Chapter 5.
11. W. Kummer, Acta. Phys. Aust. 41(1975)315.
12. Ref. 4 and J.S. Ball, private communication.
13. A.I. Alekseev, ref. 4; W.J. Shoenmaker, ref. 4.
14. H. Strubbe, Comp. Phys. Comm. 8(1974)1.
15. M.S. Milgram and H.C. Lee (to be published).
16. M.S. Milgram and H.C. Lee (in preparation).

LATTICE GAUGE THEORIES

Michael Creutz

Physics Department
Brookhaven National Laboratory
Upton, NY 11973

In the last few years lattice gauge theory has become the primary tool for the study of nonperturbative phenomena in gauge theories. The lattice serves as an ultraviolet cutoff, rendering the theory well defined and amenable to numerical and analytical work. Of course, as with any cutoff, at the end of a calculation one must consider the limit of vanishing lattice spacing in order to draw conclusions on the physical continuum limit theory. The lattice has the advantage over other regulators that it is not tied to the Feynman expansion. This opens the possibility of other approximation schemes than conventional perturbation theory. Thus Wilson used a high temperature expansion to demonstrate confinement in the strong coupling limit. Monte Carlo simulations have dominated the research in lattice gauge theory for the last four years, giving first principle calculations of nonperturbative parameters characterizing the continuum limit.

Before reviewing some of the recent results with lattice calculations, I wish to spend a few minutes reviewing the parameters of the theory of the strong interactions. First of all there are the quark masses. These presumably arise from some grand unification of the forces of nature, and are generally regarded as uncalculable in the gauge theory of the strong interactions alone. Their values are related to the pseudoscalar meson masses, which would vanish in a chirally symmetric world of vanishing quark masses. The remarkable feature of the strong interactions is that these are the only parameters. In principle all dimensionless quantities, such as the ratio of the ρ meson mass to the nucleon mass, are determined once the quark masses are given. This applies not only to mass ratios, but also to quantities such as the pion-nucleon coupling constant which once was considered a possible expansion parameter for a fundamental theory of baryons and mesons.

But what about the strong coupling constant; isn't that a parameter? Indeed, it is not; the coupling drops out of physical quantities via the phenomenon of dimensional transmutation [1], which I will now discuss in the context of the lattice regulator. On a lattice it is natural to consider a mass in units of the lattice spacing. In statistical mechanics this represents an inverse correlation length

$$\xi^{-1} = ma \quad , \qquad (1)$$

where m is the mass in question and a the lattice spacing. Now for a continuum limit we wish to take the lattice spacing to zero while holding the mass m at its physical value. Eq. (1) immediately implies that in this limit the correlation length must diverge. This is the source of the statement that in statistical mechanics one must go to a critical point to obtain a continuum limit.

For an asymptotically free theory, as considered here, we know something about how the coupling varies for the continuum limit. As the bare coupling is an effective coupling at the scale of the cutoff, it should decrease logarithmically with the cutoff [2]

$$g_0^{\,2} = \left(\gamma_0 \ln(a^{-2}\Lambda_0^{\,-2}) + (\gamma_1/\gamma_0)\ \ln\ln(a^{-2}\Lambda_0^{\,-2}) + (g_0^{\,2})\right)^{-1} \ . \qquad (2)$$

Here γ_0 and γ_1 are the usual first two terms in the Gell-Mann-Low renormalization group function. The parameter Λ_0 is an integration constant of the renormalization group equation. It will set the overall scale of the strong interactions, and will cancel from dimensionless ratios. We now take this well known dependence and solve it for the cutoff as a function of the coupling. Inserting the result in Eq. (1) shows how the correlation length diverges as we approach the critical point at vanishing coupling

$$\xi^{-1} = \frac{m}{\Lambda_0}\ (\gamma_0 g_0^{\,2})^{-\gamma_1/2\gamma_0^{\,2}}\ \exp\left(\frac{-1}{2\gamma_0 g_0^{\,2}}\right) \times \left(1 + O(g_0^{\,2})\right) \quad . \qquad (3)$$

Thus, to determine a mass in units of the integration constant Λ_0, first determine by some method such as Monte Carlo simulation the divergence of the correlation length as the coupling is reduced. The coefficient of this divergence is the desired ratio.

The important observation at this point is that Λ_0 is independent of the particular mass being measured. One could first use the correlation between operators with, say, ρ quantum numbers, and then the correlation between two proton operators. Taking the ratio of the coefficients of the corresponding divergences gives the ρ to proton mass ratio with no parameters left to adjust beyond the bare quark masses, which are already determined most naturally from the pseudo-scalar masses.

To illustrate the procedure, consider the long range linear potential between quark-like sources in the pure SU(3) gauge theory. In Fig. 1 I show Monte Carlo measurements of the effective force, in lattice units, between two such sources at various separations [3]. The points form an envelope representing the strength of the constant long range force K in units of the lattice spacing a. This force is plotted versus $\beta = 6/g_0^2$ so that the dominant exponential behavior in Eq. (3) appears as a nearly straight line when plotted on this logarithmic paper. The normalization gives the string tension in units of the square of the Λ parameter, the band plotted in the figure representing

$$\Lambda_0 = (6 \pm 1) \times 10^{-3}\sqrt{K} \qquad (4)$$

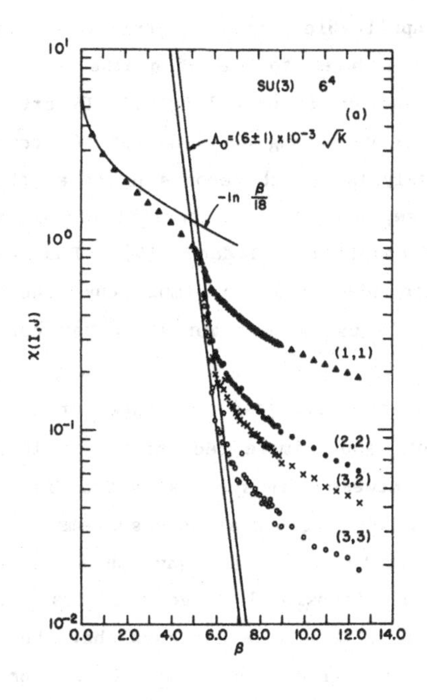

Fig. 1

The effective force between quarks in pure SU(3) lattice gauge theory.
The χ ratios are defined in Ref. [3]. Their envelope is the coefficient of the long range force, measured in lattice units.

Correcting for scheme dependence, and putting in the phenomenological value $\sqrt{K} \approx 400$ (MeV), this corresponds to the more conventional Λ_{MOM} being about 180 MeV.

In this way one can attempt to calculate from first principles the physical observables of the strong interactions. Note that the essential singularity of Eq. (3) is inherently nonperturbative. As mentioned above, this is one of the prime motivations for the lattice cutoff. In lattice gauge theory the variables are elements of the gauge group and are usually associated with the bonds of a hypercubical lattice. These represent the familiar path ordered exponentials between the corresponding sites.

Several techniques have been developed for calculations in lattice gauge theory. As the lattice is just a cutoff, one could proceed with ordinary perturbation theory. This should reproduce all the standard results as obtained with other regulators. The lattice propagators are, however, rather complicated, and thus only a few one loop calculations have been done [4]. As other schemes for perturbative analysis are rather highly developed, the value of the lattice lies elsewhere.

In his initial work on the subject, Wilson studied the strong coupling limit of the theory [5]. High temperature expansions from statistical mechanics are directly applicable here, and confinement is readily derived. Indeed, the theory reduces to one of quarks on the ends of thin strings with a finite energy per unit length. Unfortunately, as discussed above, we must leave strong coupling for the continuum limit. Mean field theory analysis, which becomes exact as the dimension of space-time becomes large, suggests the possibility of a deconfining phase transition as the coupling is reduced [6]. Duality arguments show that such a transition does exist in simple toy models based on discrete gauge groups [6]. Thus we must worry whether confinement persists in the continuum limit.

Before the Monte Carlo evidence, the strongest arguments for confinement came from the Migdal-Kadanoff approximate recursion relations [7]. These showed a strong analogy between spin systems in two dimensions and gauge theories in four. As two dimensional spin systems with a non-Abelian symmetry are believed not to have a spin wave phase, this suggests that the corresponding four dimensional gauge theories are confining for all values of coupling. Although we now know that these approximations to the renormalization group functions have some shortcomings, they have had considerable success in predicting gross phase structure and should receive more attention [8]. A recent paper of Tomboulis [9] argues that these relations provide a lower bound to the interquark potential.

This brings me to Monte Carlo simulation, which has become by far the most popular calculational tool in this field [10]. To the particle physicist, Monte Carlo provides a technique for the numerical evaluation of path integrals. The method converges reasonably well in all domains of coupling and in principle permits the evaluation of any desired correlation. The calculations in practice, however, do have inherent limitations. First of all, the lattices being four dimensional are necessarily rather limited in size, typically of order 10 sites on a side. Thus both finite volume and finite lattice spacing effects must be carefully monitored and traded off against each other. Nevertheless, the experimental observation of precocious scaling in deeply inelastic leptonic scattering suggests that a few qualitative results are possible. Secondly, statistical errors decrease only with the square root of the computer time. In particular, this has plagued the analysis of glueball masses.

Monte Carlo calculations have given two rather clear and uncontroversial numbers. The first of these is the ratio of the string tension to the asymptotic freedom scale mentioned above. The second number is the temperature of the physical transition where confinement loses meaning because the vacuum becomes a soup of gluonic flux [11]. Let me note in passing that this transition appears to be quite abrupt, probably first order, with a nearly empty vacuum going suddenly to an essentially free gas of quarks and gluons. This abruptness may be related to the successes of the bag model [12], which would suggest a transition at the temperature where the pressure of a free gluon gas equals the bag constant. The estimated deconfinement transition temperature of 180 MeV would then give a bag constant of

$$B = \frac{8}{3}\,\sigma\, T_c{}^4 = (8\pi^2 T_c{}^4/45) \approx (210\ \text{MeV})^4 \quad . \tag{5}$$

Another observable that has received considerable attention is the glueball mass. Indeed, a whole spectrum of states is being investigated, primarily by the two groups in Ref. [13]. Schierholz's talk at this conference reviews the status of these calculations. Let me only remark that the lightest 0^{++} state is reasonably uncontroversially found at 700-1000 MeV, and various methods indicate an extremely rich spectrum lies below 2 GeV.

How to include quarks in the lattice calculation remains an active and unsettled question. From an analytic point of view, a fermionic path integral is perfectly well defined; however, it is not an integral in the classical sense and thus it is unclear how to use importance sampling techniques for numerical work. Analytically removing the fermionic fields leaves an ordinary integral over the gauge fields but the weight includes the determinant of an enormous matrix. The Monte Carlo method needs numerical estimates for the changes in this determinant as the gauge fields are altered. This requires knowledge of its inverse, and several schemes for this evaluation have been proposed. In an early paper, Weingarten and Petcher [14] proposed directly evaluating the inverse with a Gauss-Seidel procedure. This involves rather intensive computation, although they were able to carry out calculations on a 2^4 site lattice. Fucito, Marinari, Parisi and Rebbi [15] proposed using Monte Carlo with auxiliary scalar fields to find the required inverse stochastically. Recent calculations [16] with this method are encouraging, indicating that the effects of the fermionic loops are controllable. Kuti [17] has recently proposed the use of a stochastic method of Von Neuman and Ulam for the inverse. Results from this technique with gauge theories are eagerly anticipated. As an estimate of the effects of the virtual quark loops, Ref. [18] has compared calculations with -2 and 0 flavors, the former obtained from a conventional Monte Carlo simulation with commuting fields. Their results indicate that hadronic masses are fairly stable, but that decay constants can receive substantial corrections.

The most publicized results with quark fields involve the "valence" or "quenched" approximation, wherein the quark propagators are calculated in a gauge field configuration obtained without any feedback from the quarks [19]. This amounts to neglecting diagrams containing internal fermionic loops. The results support the existence of chiral symmetry breaking and give a qualitative hadronic spectrum. There have, however, been some indications that systematic effects have been underestimated. See in particular Schierholz's talk at this conference.

At this point let me change the subject somewhat and discuss a few alternative simulation techniques which may have certain advantages over the usual procedures. In Ref. [20] U(1) lattice gauge theory was studied with an analog of a molecular dynamics calculation. In addition to the gauge fields U_{ij} on the lattice links, the authors also introduced conjugate momenta ℓ_{ij} and studied the classical evolution in a new time dimension of the four dimensional lattice with Hamiltonian

$$H = S(U) + 1/2 \sum_{\{i,j\}} \ell_{ij}^2 \quad . \tag{6}$$

This sets up a meandering through phase space on the surface of constant total "energy". If the behavior is ergodic, this will sample the microcanonical ensemble. Note that the procedure contains neither random numbers nor a temperature parameter. The randomness arises from the complexity of the system itself, and the temperature can be measured from the average kinetic energy

$$1/2\ kT = 1/2\ \langle \ell_{ij}^2 \rangle \quad . \tag{7}$$

In simple cases one can dispense with the extra variables ℓ_{ij} and set up a random walk through configurations of a given total action. This would represent a hybrid of the microcanonical and Monte Carlo methods. Consider for example SU(2) lattice gauge theory. When looking at a given link, the action would be unchanged if that variable were replaced with another SU(2) matrix on a spherical submanifold of the full SU(2) group. This suggests a simulation algorithm wherein one sweeps through lattice and replaces each group element with another from the corresponding sphere. This has the advantage over the differential equation approach that it allows large changes in the elements. Empirically, the algorithm seems to converge comparably to a well optimized conventional Monte Carlo approach. It has the disadvantage that extracting the temperature is not straightforward. This is not a problem for the particle theorist because the bare coupling is not a physical observable. To test scaling laws he can use any quantity which perturbatively begins as the coupling squared, for example the internal energy, which is exactly constant in this algorithm.

This procedure takes special advantage of the shape of the SU(2) manifold; indeed, generalization to other groups is unclear. I will now present a new microcanonical simulation technique which is easily applied to any type of variable [21]. It again consists of a random walk through configuration space while maintaining a constraint on the total energy. A single extra degree of freedom, a "demon", travels around the system, transferring energy as he changes the dynamical variables. This demon is analogous to the kinetic energy in the molecular dynamics method, except that he is not associated with any particular lattice variable. This demon carries a sack of energy with non negative contents. Upon visiting a variable, he attempts to change it as in a usual simulation. Instead of accepting or rejecting the change based on random numbers, he makes the change if he has sufficient energy in his sack. The latter is adjusted so that the demon's energy plus that of the lattice system remains unchanged. As the demon is essentially a free variable, his average energy is a measure of the temperature

$$\langle E_D \rangle = 1/\beta \tag{8}$$

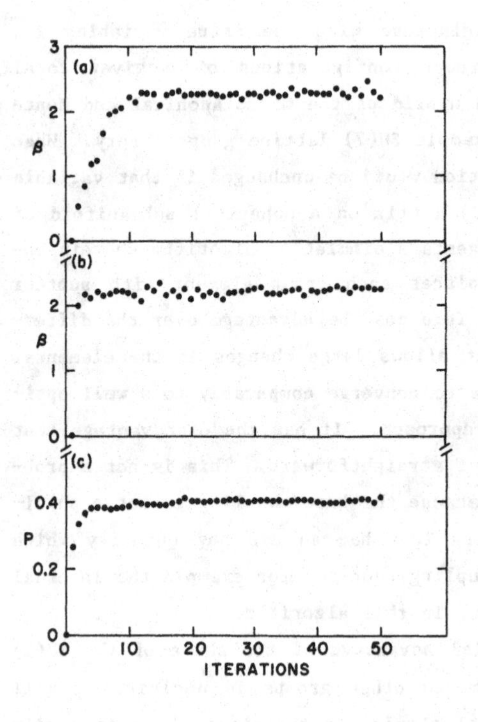

Fig. 2

Three simulations of SU(2) lattice gauge theory on a 6^4 lattice.
In (a) a demon with a large initial energy store moves sequentially through an initially ordered lattice.
In (b) he hops randomly through the system.
For commparison, (c) shows a conventional simulation at $\beta = 2.25$.

The demon might be regarded has a heat bath, but one with a very small heat capacity. If instead of a single demon, one releases a whole battalion, then they can contain an appreciable amount of energy. When the number of demons becomes large compared to the number of lattice degrees of freedom, the simulation reduces to the conventional Metropolis et al. procedure. Fig. 2 shows the results of two SU(2) simulations with this procedure and one conventional simulation.

This algorithm has some advantages. First, the demon has no need for transcendental functions; his energy becomes automatically exponentially distributed. Secondly, he is rather lenient in his demands for quality random numbers. In the case of a simulation of Ising variables, no random numbers are needed at all; the lattice uses its complexity to generate its own. Third, for discrete groups all arithmetic can be done with small integers. This means that several demons could ride on one computer word and all processing of several spins can occur simultaneously. Finally, the method does not treat the Boltzmann weight as a probability; indeed, temperature does not even appear as a parameter. As one of the problems with simulating fermionic systems is the lack of a probability interpretation for the path integral, perhaps there is a hint here.

To conclude this talk, let me observe that Monte Carlo techniques for gauge theories seem to be rapidly reaching technological limits. The glueball and fermionic calculations are using hundreds of hours on state-of-the-art computers. In contrast, note that the analytic techniques for lattice gauge theory have been relatively neglected. Whereas a few tens of particle physicists have actively pursued strong coupling expansions or Migdal-Kadanoff recursion relations, an order of magnitude more have worked on the numerical simulations. This suggests that more balance and perhaps hybrid techniques are in order. In addition, I feel that we are a long way from the last word on techniques for dealing with fermions.

ACKNOWLEDGMENT

This work has been authored under DOE Contract #DE-AC02-76CH00016.

REFERENCES

[1] S. Coleman and E. Weinberg, Phys. Rev. D7 (1973) 1888.

[2] D. Gross and F. Wilczek, Phys. Rev. Lett. 30 (1973) 1343; Phys. Rev. D8 (1973) 3633;
W.E. Caswell, Phys. Rev. Lett. 33 (1974) 244;
D.R.T. Jones, Nucl. Phys. B75 (1974) 531.

[3] M. Creutz and K.J.M. Moriarty, Phys. Rev. D26 (1982) 2166.

[4] A. Hasenfratz and P. Hasenfratz, Phys. Lett. 93B (1980) 165.

[5] K. Wilson, Phys. Rev. D10 (1974) 2445.

[6] R. Balian, J.M. Drouffe, C. Itzykson, Phys. Rev. D10 (1974) 3376; D11 (1975) 2098; D11 (1975) 2104.

[7] L.P. Kadanoff, Rev. Mod. Phys. 49 (1977) 267.

[8] K.M. Bitar, S. Gottlieb, C. Zachos, Phys. Rev. D26 (1982) 2853.

[9] E. Tomboulis, Phys. Rev Lett. 50 (1983) 885.

[10] M. Creutz, L. Jacobs, C. Rebbi, Physics Reports (in press).

[11] L. McClerran and B. Svetitsky, Phys. Lett. 98B (1981) 195;
J. Kuti, J. Polonyi, K. Szlachanyi, Phys. Lett. 98B (1981) 199;
K. Kajantie, C. Montonen, E. Pietarinen, Zeit. Phys. C9 (1981) 253;
J. Engels, F. Karsch, H. Satz, J. Montvay, Phys. Lett. 101B (1981) 89; Nucl. Phys. B205 (1982) 545.

[12] K. Johnson, talk at this conference.

[13] B. Berg and A. Billoire, Phys. Lett. 114B (1982) 324;
K. Ishikawa, G. Schierholz, M. Teper, Phys. Lett. 110B (1982) 399.

[14] D. Weingarten and D. Petcher, Phys. Lett. 99B (1981) 333.

[15] F. Fucito, E. Marinari, G. Parisi, C. Rebbi, Nucl. Phys. B180 (1981) 369.

[16] H. Hamber, E. Marinari, G. Parisi, C. Rebbi, preprint (1983).

[17] J. Kuti, Phys. Rev. Lett. 49 (1982) 183.

[18] W. Duffy, G. Guralnik, D. Weingarten, preprint (1983)

[19] D. Weingarten, Phys. Lett. 109B (1982) 57;
H. Hamber and G. Parisi, Phys. Rev. Lett. 47 (1982) 1792;
E. Marinari, G. Parisi, C. Rebbi, Phys. Rev. Lett. 47 (1981) 1798.

[20] D. Callaway and A. Rahman, Phys. Rev. Lett. 49 (1982) 613.

[21] M. Creutz, preprint (1983).

FINITE ELEMENT APPROXIMATION IN QUANTUM THEORY

Carl M. Bender
Department of Physics
Washington University
St. Louis, Missouri 63130

I. INTRODUCTION

Many of the approaches to numerical quantum field theory which have been developed to date start from the Euclidean path integral formulation of the theory. This is because it is possible, using path integrals, to write closed form, albeit formal and difficult to evaluate, expressions for the quantities of physical interest, namely the Green's functions. Thus, in this approach, the focus is not on finding the direct solution to the equations of quantum field theory, but rather on developing an effective and reliable method for evaluating the functional integrals representing the Green's functions.

The novelty of the present work [1] is that we have been able to use the method of finite elements, a technique widely employed to solve numerically partial differential equations arising in classical continuum mechanics and fluid dynamics [2], to solve operator quantum field equations in Minkowski space directly. An accurate evaluation of a functional integral requires a great many Monte Carlo passes through the lattice. In our approach, we do operator time-stepping, which requires only one pass through the lattice. We believe that the finite element method is advantageous in quantum field theory because, unlike other methods, it treats bosons and fermions on an equal footing. This is possible because in this method one does not introduce a finite difference approximation for derivatives.

II. DESCRIPTION OF THE FINITE ELEMENT METHOD

The finite element method was developed for the numerical solution of ordinary or partial differential equations. In this method one begins by partitioning the domain D on which the equation is to be solved into a collection of nonoverlapping patches, called finite elements. The finite elements are usually chosen to be polygons, whose size and shape can vary over the domain in accord with the requirements of a particular problem. On each finite element, the unknown function is represented by a low order polynomial, whose coefficients are determined by two types of conditions. First, one wants to ensure that the local approximations to the function on each finite element can be pieced together consistently to give a global representation of the function. To do this, one requires the function, and sometimes its derivatives, to be continuous across the boundary of contiguous finite elements. For those finite elements which have a boundary coinciding with the boundary of the domain D, one must also impose the boundary conditions associated with the differential operator. The second set of conditions results from imposing the differential equation; this is often done by rewriting it in variational form. These two conditions give a set of equations which determine the coefficients of the polynomials on each finite element and hence an approximate solution to the differential equation.

We illustrate the finite element method by two very elementary ordinary differential equation examples.

Example 1

Given $y'(x) = y(x)$ and $y(0) = 1$, find $y(1)$. The exact solution of this problem is $y(1) = e = 2.718281828...$. How well can we compute e using the method of finite elements? We decompose the interval $0 \leq x \leq 1$ into N intervals of length 1/N. On the ith interval we represent $y(x)$ by $a_i + b_i x$.

Since the differential equation is first order we impose only the constraint of continuity at the boundaries of the intervals. We clearly cannot impose the differential equation throughout each interval, so we impose it at just one point on every interval which we choose to be the center of the interval.

Suppose there is just one finite element ($N = 1$). Then for $0 \leq x \leq 1$, $y(x) = a_1 + b_1 x$. We impose the boundary condition $y(0) = 1$ by taking $a_1 = 1$. We impose the differential equation at $x = 1/2$, the center of the interval. This gives $b_1 = 2$. Thus, we have

$$y(x) = 1 + 2x \quad , \qquad 0 \le x \le 1 \quad ,$$

and we predict $y(1) = 3$ $(N = 1)$, with a remarkably small relative error of 10%!

Now suppose there are two elements. Imposing $y(0) = 1$, continuity at $x = 1/2$, and the differential equation at $x = 1/4$ and $x = 3/4$, the center of each interval, we obtain $y(1) = 25/9 = 2.7777....$ $(N = 2)$ which differs from the exact answer by relative error of less than 2%!

This problem is sufficiently simple that we can obtain $y(1)$ in closed form for any number N of finite elements. The result is

$$y(1)(\text{N elements}) = \left(\frac{2N + 1}{2N - 1}\right)^N \quad .$$

For large N this expression rapidly approaches the exact answer e:

$$y(1)(\text{N elements}) \sim e\left(1 + \frac{1}{12N^2}\right) \quad , \quad (N \to \infty) \quad .$$

The relative error,

$$\frac{y(1)(\text{N elements}) - y(1)(\text{exact})}{y(1)(\text{exact})} = \frac{1}{12N^2} \quad ,$$

rapidly decays to zero as N increases. For example, when $N = 10$ the relative error is less than 1 part in a thousand.

Richardson extrapolation can be used to improve the relative error by many orders of magnitude.

<u>Example 2</u>

Consider the eigenvalue problem $y''(x) = Ey(x) = 0$, $y(0) = y(1) = 0$. The exact answer for the smallest eigenvalue is of course $E = \pi^2$ $= 9.8696043...$. There are two ways to attack this problem using the method of finite elements:

<u>Method 1:</u> Using the same kind of approach as that in Example 1, we decompose the interval $0 \le x \le 1$ into N intervals of length 1/N. On the <u>i</u>th interval we approximate $y(x)$ by the <u>quadratic</u> polynomial

$$y_i(x) = a_i + b_i x + c_i x^2$$

[We demand that the approximation to $y(x)$ be continuously differentiable because the differential equation is second order. This would not be possible if the finite elements were linear polynomials.] Next we impose the differential equation at the center of each interval and the boundary conditions $y(0) = y(1) = 0$. The result is a system of N homogeneous linear equations which, by Cramer's rule, have a nonzero solution only if the determinant of the coefficients vanishes. This gives a polynomial equation for E having N roots. These roots closely approximate the first N eigenvalues of the eigenvalue problem under consideration.

For example, when $N = 1$, we approximate $y(x)$ by $y_1(x) = a + bx + cx^2$. Imposing the boundary conditions $y(0) = y(1) = 0$ determines all but one of the parameters, giving $y_1(x) = bx - bx^2$. Next we impose the differential equation at $x = 1/2$:

$$y''\left(\tfrac{1}{2}\right) + Ey\left(\tfrac{1}{2}\right) = -2b + Eb/4 = 0$$

This equation has a nonzero solution for b if $E = 8$.

In Table 1 we list eleven approximants obtained in this manner to the smallest eigenvalue. Observe that this sequence rapidly converges to the exact value π^2. The eleventh approximant has a relative error of 1/3%.

Table 1
Finite element approximations to the smallest eigenvalue E of the equation $y''(x) + Ey(x) = 0$, $y(0) = y(1) = 0$. E_N is the approximation to E obtained using N finite elements, each finite element being a quadratic polynomial. The exact answer is $\pi^2 = 9.8696043...$.

N	E_N
1	8
2	32/3 = 10.667
3	72/7 = 10.286
4	10.113
5	10.028
6	9.980
7	9.951
8	9.932
9	9.919
10	9.910
11	9.903

Method 2: A slightly fancier approach begins by restating the eigenvalue problem in variational form. The smallest eigenvalue E is the minimum value of

$$I[y] = \frac{\int_0^1 [y'(x)]^2 \, dx}{\int_0^1 [y(x)]^2 \, dx} \quad , \quad y(0) = y(1) = 0 \quad .$$

We now minimize this functional on the space of functions $y(x)$ represented as polynomials on finite elements and vanishing at $x = 0$ and $x = 1$. Note that since only one derivative of y appears in the functional, it is sufficient to use linear polynomials; $y(x)$ need no longer be continuously differentiable as in Method 1.

The technique consists of choosing a value N, patching the N polynomials together to make the function $y(x)$ continuous, satisfying the boundary conditions at $x = 0$ and $x = 1$, and evaluating $I[y]$ in terms of the remaining undetermined polynomial coefficients. Finally, we differentiate with respect to each of these coefficients to obtain the minimum of $I[y]$. Doing this for each value of N produces a sequence of approximations I_N which rapidly and monotonically approaches the lowest eigenvalue π^2 from above. (The sequence is monotone decreasing because the problem is in variational form.)

Table 2
The first eight finite element approximations to the smallest eigenvalue E of the equation $y''(x) + Ey(x) = 0$, $y(0) = y(1) = 0$. I_N is the approximation to E obtained using N quadratic finite elements to minimize $I(y) = \int_0^1 [y'(x)]^2 \, dx / \int_0^1 [y(x)]^2 \, dx$. The exact answer is $\pi^2 = 9.86960430...$

N	I_N
1	10.0
2	10.0
3	9.89011
4	9.87553
5	9.87192
6	9.87070
7	9.87018
8	9.86994

If we choose to use linear polynomials on the finite elements then $I_2 = 12$, $I_3 = 10.8$, and so on (the result is ambiguous for $N = 1$). The $N = 3$ result has a relative error of less than 10%. If we choose to use quadratic polynomials the numbers improve dramatically. Eight finite elements give I_N correct to 1 part in 3×10^4. See Table 2.

The conventional methods of finite elements have often been applied to differential equation problems in classical physics involving fluid mechanics and solid mechanics. However, now the coefficients of the polynomials on the finite elements are operators, whose properties are determined by the equal time commutation relations.

III. THE FINITE ELEMENT APPROXIMATION IN QUANTUM MECHANICS

To illustrate our procedure we consider the problem of solving the Heisenberg equations of motion for a one-dimensional quantum system. If the Hamiltonian is

$$H = p^2/2 + V(q) \quad , \tag{1}$$

The Heisenberg equations are

$$dq(t)/dt = p(t) \quad , \quad dp(t)/dt = f[q(t)] \quad . \tag{2}$$

where $f(q) = -V'(q)$. The quantum mechanical problem consists of solving (2) for the operators $p(t)$ and $q(t)$ given the equal time commutation relation

$$[q(t),p(t)] = i \quad . \tag{3}$$

We solve this problem first on a single finite element. We approximate $q(t)$ and $p(t)$ by linear functions of t:

$$q(t) = \left(1 - \frac{t}{h}\right)q_0 + \frac{t}{h}q_1 \quad , \quad p(t) = \left(1 - \frac{t}{h}\right)p_0 + \frac{t}{h}p_1 \, , \; 0 < t < h \quad . \tag{4}$$

Substituting (4) into (2) and evaluating the result at the <u>center</u> of the time interval $t_0 = h/2$ gives

$$\frac{q_1 - q_0}{h} = \frac{p_0 + p_1}{2} \quad , \tag{5}$$

$$\frac{p_1 - p_0}{h} = f\left(\frac{q_0 + q_1}{2}\right) \quad . \tag{6}$$

Equations (5) and (6) are a discrete form of the equations of quantum mechanics. But, are they consistent with the equal time commutation relation (2)? At $t=0$, (3) reads $[q_0,p_0]=i$. The question is, if we solve (5) and (6) simultaneously for the operators p_1 and q_1, will the commutator $[q_1,p_1]$ also have the value i? We can evaluate $[q_1,p_1]$ for any function f: Commute (5) on the right with $p_0 + p_1$,

$$[q_1 - q_0, p_0 + p_1] = 0 \quad , \tag{7}$$

and commute (6) on the left with $q_0 + q_1$,

$$[q_0 + q_1, p_1 - p_0] = 0 \quad . \tag{8}$$

Adding (7) and (8) gives

$$[q_1, p_1] = [q_0, p_0] = i \quad . \tag{9}$$

Thus, we have proved that the method of finite elements <u>is</u> consistent with the equal time commutation relation on a single finite element. But clearly, for a collection of N elements this argument may be used iteratively to show that the difference equations in (5) and (6) are consistent with (3) at the endpoints of each of the finite elements.

We make two observations. First, the equal time commutation relation holds only at the endpoints of a finite element, but not at any interior point. Second, the result in (9) depends crucially upon our having imposed the Heisenberg equations in (2) at the <u>center</u> of the finite element, $t=t_0=h/2$. For any other value of t_0, (9) is false. Thus the operator properties of quantum mechanics uniquely determine the value of t_0. In a classical problem, there is no such constraint on the value of t_0.

Even though (5) and (6) are operator equations, we can solve them explicitly for p_1 and q_1, in terms of p_0 and q_0. First, we solve (5) for p_1:

$$p_1 = \frac{2}{h}\left(q_1 - q_0\right) - p_0 \quad . \tag{10}$$

and use this result to eliminate p_1 from (6):

$$-\frac{2}{h}\, p_0 - \frac{4}{h^2}\, q_0 = g\left(\frac{q_1 + q_0}{2}\right) \quad , \tag{11}$$

where $g(x) = f(x) - 4x/h^2$ is the function which completely characterizes the dynamical content of this quantum theory. The solution to (11) is given in terms of g^{-1}, where $g^{-1}(y)$ is the solution to the classical equation $y = g(x)$. From (11) we have

$$q_1 = - q_0 + 2g^{-1}\left(- \frac{2}{h} p_0 - \frac{4}{h^2} q_0\right) \tag{12}$$

and from (10) we obtain

$$p_1 = -p_0 - \frac{4}{h} q_0 + \frac{4}{h} g^{-1}\left(- \frac{2}{h} p_0 - \frac{4}{h^2} q_0\right) \quad . \tag{13}$$

The result in (12) and (13) is the one time-step solution of the quantum-mechanical initial value problem in (2) [3]. We generalize to N time steps (N finite elements) by iterating (12) and (13) N times to express p_N and q_N in terms of p_0 and q_0. Note that the solution takes the form of a continued (nested) function $A + g^{-1}(A + g^{-1}(A + g^{-1}(A + \ldots)))$. If $[q_1, p_1]$ had not turned out to be i, but had differed from i by a (presumably) small q-number part that vanishes with the spacing h, then we would not be able to obtain the N-finite-element solution by iteration, and indeed we would not even be able to solve for p_2 and q_2.

IV. SOME NUMERICAL RESULTS

For the harmonic oscillator defined by $H = p^2/2 + m^2q^2/2$ the operator equations (5) and (6) are linear and the solution in (12) and (13) can be written in matrix form:

$$\begin{pmatrix} p_1 \\ q_1 \end{pmatrix} = M \begin{pmatrix} p_0 \\ q_0 \end{pmatrix} \quad .$$

Although M is not symmetric

$$M = \frac{1}{1 + m^2h^2/4} \begin{bmatrix} 1 - h^2m^2/4 & -hm^2 \\ h & 1 - h^2m^2/4 \end{bmatrix} \quad , \tag{14}$$

it can be written as a similarity transform of a diagonal matrix $D, M = QDQ^{-1}$, whose entries are

$$d_{11}, d_{12} = (1-h^2m^2/4 \pm ihm)/(1+m^2h^2/4) = \exp\{\pm i \text{ arc sin}[mh/(1+m^2h^2/4)]\}. \tag{15}$$

The solution for p_n and q_n is

$$p_n = A(d_{11})^n + B(d_{22})^n \tag{16}$$

and

$$q_n = C(d_{11})^n + D(d_{22})^n \tag{17}$$

where $A = (p_0 + imq_0)/2$, $B = (p_0 - imq_0)/2$, $C = A/(im)$, $D = iB/m$. Observe that in the continuum limit $nh = T$, $mh \to 0$, the diagonal entries d_{11}, d_{22} become

$$d_{11}, d_{22} = e^{\pm imT} \quad . \tag{18}$$

Thus, the fields p and q have a time oscillation given by (18) and we can read off the exact answer for the energy gap ΔE (the excitation of the first energy level over the vacuum)

$$\Delta E = E_1 - E_0 = m \quad . \tag{19}$$

We can also compute the two-point function, which in the continuum is defined by

$$G(t,t') \equiv \langle 0|T[q(t)q(t')]|0\rangle \quad . \tag{20}$$

$G(t,t')$ satisfies the differential equation

$$\frac{\partial^2 G}{\partial t^2} + m^2 G = -i\delta(t-t') \quad ,$$

whose solution is

$$G(t,t') = \frac{1}{2m} \exp(-im|t-t'|) \ . \tag{21}$$

To compute G on the lattice we substitute (17) into the formula

$$\begin{aligned} G_{NM} &= \langle 0|\theta(N-M)q_N q_M + q_M q_N \theta(M-N)|0\rangle \\ &= \frac{1}{2m}\left(d_{22}^N\, d_{11}^M\, \theta_{N-M} + d_{22}^M\, d_{11}^N\, \theta_{M-N}\right) \\ &= \frac{1}{2m}\, e^{-i|N-M| \arcsin[mh/(1+m^2h^2/4)]} \quad . \end{aligned} \tag{22}$$

In the continuum limit $T = t-t'$, $mh \to 0$, $t = Nh$, $t' = Mh$, G_{NM} in (22) becomes $G(t,t')$ in (21).

For the anharmonic oscillator, whose Hamiltonian is $H = p^2/2 + \lambda q^4/4$, the energy gap is known to be $\Delta E = E_1 - E_0 = (1.08845)\lambda^{1/3}$. We can calculate E accurately from (5), (δ), (9) and the equation

$$\langle n|p|m\rangle = i(E_n - E_m)\langle n|q|m\rangle \quad ,$$

where $|n\rangle$ is an eigenstate of the Hamiltonian H with energy E_n, using a variational procedure in which we include two states and one time step, three states and two time steps, and so on. Using just two states we obtain $\Delta E = (\lambda/2)^{1/3} = (0.79370)\lambda^{1/3}$ which is 27% off. Including three states gives $\Delta E = (3\lambda/2)^{1/3} = (1.14471)\lambda^{1/3}$ which is 5% off. By going from one to two time steps the error goes down by a factor of five, just as in the simple calculation of e in Example 1.

V. HIGHER-DIMENSIONAL QUANTUM FIELD THEORY

Now we show how to generalize this method to quantum field theory. Consider a scalar field theory in two-dimensional Minkowski space. We write the operator field equations as a coupled first-order system so that we can continue to use linear approximations to the fields on finite elements

$$\pi = \phi_t \quad , \quad \gamma = \phi_x \quad , \quad \pi_t - \gamma_x + f(\phi) = 0 \quad . \tag{23}$$

We introduce rectangular finite elements whose length in the time direction is h and in the space direction is k. On the m, n-element, the field ϕ is approximated by the bilinear polynomial

$$\begin{aligned}\phi(x,t) = &\left(1-\frac{t}{h}\right)\left(1-\frac{x}{k}\right)\phi_{m-1,n-1} + \left(1-\frac{t}{h}\right)\left(\frac{x}{k}\right)\phi_{m,n-1} \\ &+ \left(\frac{t}{h}\right)\left(1-\frac{x}{k}\right)\phi_{m-1,n} + \left(\frac{t}{k}\right)\left(\frac{x}{k}\right)\phi_{m,n}\end{aligned} \tag{24}$$

where the coefficient $\phi_{m,n}$ is the value of the field operator at the site (m,n). The fields π and γ are represented in a similar way.

Our objective is to show how to advance one step in the time direction. We consider a single horizontal row of M finite elements and impose (23). The result is the following system of difference equations.

$$\frac{1}{4}(\pi_{m-1,0}+\pi_{m,0}+\pi_{m-1,1}+\pi_{m,1})=\frac{1}{2h}(\phi_{m,1}+\phi_{m-1,1}-\phi_{m,0}-\phi_{m-1,0}) ,$$

$$\frac{1}{4}(\gamma_{m-1,0}+\gamma_{m,0}+\gamma_{m-1,1}+\gamma_{m,1})=\frac{1}{2k}(\phi_{m,1}+\phi_{m,0}-\phi_{m-1,1}-\phi_{m-1,0}) ,$$

$$\frac{1}{2h}(\pi_{m,1}+\pi_{m-1,1}-\pi_{m,0}-\pi_{m-1,0})-\frac{1}{2k}(\gamma_{m,1}+\gamma_{m,0}-\gamma_{m-1,1}-\gamma_{m-1,0})$$
$$= f\left[\frac{1}{4}(\phi_{m-1,0}+\phi_{m,0}+\phi_{m-1,1}+\phi_{m,1})\right] . \tag{25}$$

$m = 1,2,\ldots M$. We take each finite element for ϕ and π to represent one degree of freedom and define the dynamical variables

$$\Phi_{m,n}=\frac{1}{2}(\phi_{m,n}+\phi_{m-1,n}) , \quad \Pi_{mn}=\frac{1}{2}(\pi_{m,n}+\pi_{m-1,n}) . \tag{26}$$

If we eliminate $\gamma_{m,n}$ from (25) and express the resulting equations in terms of the dynamical variables we obtain a system of 2M equations whose general structure is

$$\Pi_{m,0}+\Pi_{m,1}=\frac{2}{h}(\Phi_{m,1}-\Phi_{m,0}) , \tag{27}$$

$$\Pi_{m,1}-\Pi_{m,0}=\sum_{g=1}^{M} S_{m,j}(\Phi_{j,1}+\Phi_{j,0})+F(\Phi_{m,1}+\Phi_{m,0}) \tag{28}$$

$m = 1,2,3,\ldots M$. Here S is a symmetric matrix [4] and F is a nonlinear function simply related to f in (15).

When written in terms of $\Phi_{m,n}$ and $\Pi_{m,n}$, the ETCR's for the fields, $[\phi(x,t),\pi(y,t)] = i\delta(x-y)$, become

$$[\Phi_{j,n},\Phi_{\ell,n}]=0 , \quad [\Pi_{j,m},\Pi_{\ell,m}]=0 , \quad [\Phi_{j,m},\Pi_{\ell,m}]=\frac{i}{k}\delta_{j,\ell} . \tag{29}$$

The consistency problem here is to show that if (29) holds for $n=0$ then by virtue of (27) and (28) it also holds for $n=1$.

The proof of consistency is not simple. There are three steps. First, we eliminate $\Pi_{m,1}$, $m = 1,2,3,\ldots M$, from (28) using (27). Thus (28) takes the form

$$\Pi_{m,0}+\frac{2}{h}\Phi_{m,0}=\sum_{j=1}^{M} S_{m,j}(\Phi_{j,1}+\Phi_{j,0})+G(\Phi_{m,1}+\Phi_{m,0}), \quad j=1,2,\ldots M . \tag{30}$$

where G is simply related to F. Because

$$[\Pi_{j,0} + \frac{2}{h}\Phi_{j,0}, \Pi_{\ell,0} + \frac{2}{h}\Phi_{\ell,0}] = 0 \quad ,$$

we can in principle solve (30) for $\Phi_{m,1} + \Phi_{m,0}$ in terms of $\Pi_{m,0} + \frac{2}{h}\Phi_{m,0}$. It follows from (27) at $n = 0$ that

$$[\Phi_{j,1} + \Phi_{j,0}, \Phi_{\ell,1} + \Phi_{\ell,0}] = 0 \quad . \tag{31}$$

Second, we replace G in (30) by εG, where ε is a small parameter. Then, assuming that G has a Taylor series, we solve for $\Phi_{m,1}$ as a perturbation series in powers of ε. We can show that to all orders in powers of ε,

$$[\Phi_{j,1}, \Phi_{\ell,1}] = 0 \quad , \tag{32}$$

so long as S is a symmetric matrix. This is the difficult part of the proof.

Third, combining (31) and (32) gives

$$[\Phi_{j,1} - \Phi_{j,0}, \Phi_{\ell,1} - \Phi_{\ell,0}] = 0 \quad . \tag{33}$$

We complete the proof by using the same procedure as that leading to (9): we commute (27) and (28) with $\Phi_{j,0} \pm \Phi_{j,1}$ and add the resulting commutators to establish (29) for $n = 1$. It is quite interesting that this proof depends critically on the symmetry of the matrix S; the result is not sensitive to the choice of the nonlinear function f except for the assumption that it has a Taylor series expansion. Having established the consistency for the first time step, by induction (29) holds for all values of n. This proof also shows how to solve the operator equations (27) and (28) algebraically.

Currently K. Milton, D. Sharp, and the author are studying the problem of applying the finite element techniques to fermi systems. We find that such systems are treated consistently by the discretization described in this paper. Our ultimate goal is to compute from the difference equations like (27) and (28) objects such as unequal-time commutators, light-cone commutators and S-matrix elements. We are working on these computations in various models.

The author is grateful to the U.S. Department of Energy for financial support.

REFERENCES

[1] This work was done in collaboration with D. H. Sharp. See C. M. Bender and D. H. Sharp, Phys. Rev. Lett. 50, 1535 (1983).

[2] Useful general references on the finite element method are G. Strang and G. J. Fix, An Analysis of The Finite Element Method, (Prentice-Hall Inc., Englewood Cliffs, 1973) and T. J. Chung, Finite Element Analysis in Fluid Dynamics, (McGraw-Hill, New York, 1978).

[3] There are several interesting remarks to be made here. One intriguing question is whether (12) and (13) might be used in combination with $[q_0, p_0] = i$ to find a spectrum generating algebra. Second, one may ask what happens when the equation $y = g(x)$ has multiple roots; that is, what role is played by instantons in these lattice calculations?

[4] The matrix S is a numerical matrix containing the lattice spacings h and k. It is symmetric because with properly chosen boundary conditions the operator ∇^2 in the continuum is symmetric.

TOPOLOGICAL CHARGE IN LATTICE GAUGE THEORY

J. Polonyi

Department of Physics, University of Illinois at Urbana-Champaign
1110 W. Green Street, Urbana, IL 61801

1. INTRODUCTION

Lattice regularization has proved to be succesful in the numerical solution of asymptotically free quantum field theories [1,2]. However it is believed that lattice regularization changes the semicalssical structure of the theories considerably. The reason is that states of the continuum theory with different topological charge are connected by singular gauge transformations [3]. With finite lattice spacing there are no singularities. Consequently one expects no tunneling in lattice theories. The aim of the present study is to show the existence of topological structure in lattice gauge theory. I give a simple expression for the topological charge of lattice gauge theories in four dimensions and I present the preliminary result of a Monte Carlo calculation of the distribution of this quantity on a small lattice in SU(2) gauge theory. The measured distribution is compatible with the dilute gas picture of certain pseudoparticles.

Luscher has already proposed a formula for topological charge in lattice gauge theories [4] but it is too complicated for practical calculations. A simple formula has been given for the topological charge density of lattice gauge fields in [5]. Unfortunately it contains perturbative contributions which one has to substract from the Monte Carlo results.

The organisation of this paper is the following. In Section 2, I show that the topological charge of a periodic gauge field configuration is given by the homotopy index of a map from the three dimensional torus into the gauge group realised by thermal Wilson lines. Section 3 contains a brief description of this homotopy index on the lattice. A rough estimation and the Monte Carlo results for the distribution of the topological charge on a small lattice are presented in Section 4. Finally the conclusion is given in Section 5.

2. TOPOLOGICAL CHARGE IN A FINITE PERIODIC BOX

Consider SU(n) gauge theory in a four dimensional finite continous space-time box $0 < x^\mu < L$. The action and the topological charge are given by

$$S = \frac{1}{4g^2} \int d^4x \, \mathrm{tr}\, F_{\mu\nu}F_{\mu\nu},$$

$$Q = \frac{-1}{32\pi^2} \int d^4x \, \varepsilon_{\mu\nu\rho\sigma} \mathrm{tr}\, F_{\mu\nu}F_{\rho\sigma}, \tag{1}$$

$$F_{\mu\nu} = \partial_\mu A_\nu - \partial_\nu A_\mu + [A_\mu, A_\nu],$$

respectively where the field A contains the generators $\sigma^i/2$ of the gauge group

$$A_\mu(x) = g \frac{\sigma^i}{2} A^i_\mu(x). \tag{2}$$

In a finite volume the dynamics does not quantise the topological charge. It is the role of the boundary condition to determine the spectrum of Q. The construction presented below gives integer topological charge for periodic field configuration except for a subset of zero measure in the configuration space. The apparent difficulty in defining topological charge in periodic space-time is that Q vanishes for sufficiently smooth configurations

$$Q = \frac{-1}{8\pi^2} \int d^4x \, \partial_\mu \varepsilon_{\mu\nu\rho\sigma} \, \mathrm{tr}\, A_\nu\left(\partial_\rho A_\sigma + \frac{2}{3} A_\rho A_\sigma\right). \tag{3}$$

But we do not exclude field configurations having gauge singularities from the path integral. These configurations may give nonzero topological charge. A reasonable procedure to define explicit gauge invariant integrals like (1) is to choose a physical gauge where the gauge singularities are transformed away. The axial gauge where $A_o = 0$ is adequate for this purpose. In this gauge we have to relax the strict periodicity in the time direction. Instead we require periodicity up to a gauge transformation described by the parameter

$$\Omega(\vec{x}) = P \exp\{i \int_o^L dx^o \; A_o(\vec{x},x^o)\}. \tag{4}$$

To fix the remaining gauge degrees of freedom I chose the following "periodic axial gauge" for the spacelike components of A [7]:

$$\begin{aligned} &\partial_3 A_3(x^1, x^2, x^3, x^o = 0) = 0, \\ &\partial_2 A_2(x^1, x^2, x^3 = x^o = 0) = 0, \\ &\partial_1 A_1(x^1, x^2 = x^3 = x^o = 0) = 0. \end{aligned} \tag{5}$$

This gauge condition leaves the strict periodic boundary condition unchanged in the space direction. The ghost field resulting from this gauge fixing procedure is decoupled from the gauge field.

For sufficiently smooth field configurations in this gauge one has [13]

$$Q = \frac{-1}{24\pi^2} \int d^3x \; \varepsilon_{ijk} \; \mathrm{tr}[\Omega\partial_i\Omega^{-1}\Omega\partial_j\Omega^{-1}\Omega\partial_k\Omega^{-1}] + \int d^3x \; \partial_i F_i. \tag{6}$$

Neglecting the surface terms we conclude that the topological charge of a periodic gauge configuration is given by the homotopy index of a map of the three dimensional torus into the gauge group.

It is worthwhile to note the relation of the topological charge at finite temperature and the free energy of a static quark given by $F = -kT \log \langle \mathrm{tr}\Omega(\vec{x})\rangle$. In the confining phase of the SU(2) theory $(T < T_c)$ $F = \infty$ and the effective potential of $\mathrm{tr}\Omega(\vec{x})$ has one minimum at $\mathrm{tr}\Omega(\vec{x}) = 0$. In the deconfining phase $(T > T_c)$ F is finite and the effective potential has two minimas at $tr\Omega(\vec{x}) = \pm a(T)$. Configurations where $tr\Omega(\vec{x}) \sim a \neq 0$ have zero topological charge. Consequently one expects suppression of the topological charge for $T > T_c$. This relation allows us to calculate F in the dilute gas approximation [6].

3. HOMOTOPY INDEX ON THE LATTICE

First I give the precise definition of the gauge described in the previous Section on the lattice. We will frequently need the one parameter subgroup connecting the identity with a given gauge group element. In order to simplify the formulas we restrict ourselves to the case where the gauge group G = SU(2). The one parameter subgroup g connecting 1 with $g(\vec{\alpha}) = \cos\alpha/2 + i\,\vec{\alpha}\vec{\sigma}/\alpha \sin\alpha/2$ is given by $g^{\tau}(\vec{\alpha}) = g(\tau\vec{\alpha})$ and it is well defined for $g \neq -1$. The gauge transformation which transforms a given lattice configuration $U_\mu(n_\nu)$, $1 < n_\nu < N$ to this gauge is

$$U'_\mu(n) = g_n U_\mu(n)\, g^{-1}_{n+\hat{\mu}} , \tag{7}$$

where the parameter g(n) of the transformation is given by

$$\begin{aligned}
g(m) &= K_1H_1K_2H_2K_3H_3H_4 \\
H_1(m) &= \Pi_{i=1}^{m_1-1}\; U_1(i, n_2 = n_3 = n_o = 1), \\
H_2(m) &= \Pi_{i=1}^{m_2-1}\; U_2(m_1,\, i,\, n_3 = n_o = 1), \\
H_3(m) &= \Pi_{i=1}^{m_3-1}\; U_3(m_1,\, m_2,\, i,\, n_o = 1), \\
H_4(m) &= \Pi_{i=1}^{m_4-1}\; U_4(m_1,\, m_2,\, m_3,\, i), \\
K_1(m) &= [\Pi_{i=1}^{N}\; U_1(i,\, n_2 = n_3 = n_o = 1)]^{m_1-1/N}, \\
K_2(m) &= [\Pi_{i=1}^{N}\; U_2(m_1,\, 2,\, n_3 = n_o = 1)]^{m_2-1/N}, \\
K_3(m) &= [\Pi_{i=1}^{N}\; U_3(m_1,\, m_2,\, i,\, n_o = 1)]^{m_3-1/N}.
\end{aligned} \tag{8}$$

The thermal Wilson line in this gauge reads

$$\Omega(\vec{n}) = g_{\vec{n},1} [\Pi_{i=1}^{N} U_o(\vec{n},i)] g_{\vec{n},1}^{-1} . \tag{9}$$

We are now in position to introduce the interpolating field for the thermal Wilson line. First we divide the three dimensional torus $n = 1$, $1 < n_\lambda < N$ on the lattice into tetrahedrons (c.f. Fig. 1). Then we define an interpolating field $w(x)$ in each tetrahedron in such a way that they match on common surfaces. A possible interpolating field is given by

$$w(\vec{\tau}) = G_2(G_2^{-1}g_4)^{\tau_3}, \quad G_2 = G_1(G_1^{-1}g_3)^{\tau_2}, \quad G_1 = g_1(g_1^{-1}g_2)^{\tau_1}$$

$$\vec{x}(\vec{\tau}) = \vec{Y}_2 + \tau_3(\vec{n}_4 - \vec{Y}_2), \quad \vec{Y}_2 = \vec{Y}_1 + \tau_2(\vec{n}_3 - \vec{Y}_1), \tag{10}$$

$$\vec{Y}_1 = \vec{n}_1 + \tau_1(\vec{n}_2 - \vec{n}_1),$$

where $\vec{n}_i$ and g_i are the vertices of the tetrahadron in space and in group space respectively.

The homotopy index of the thermal Wilson lines is simply the sum of the integrals

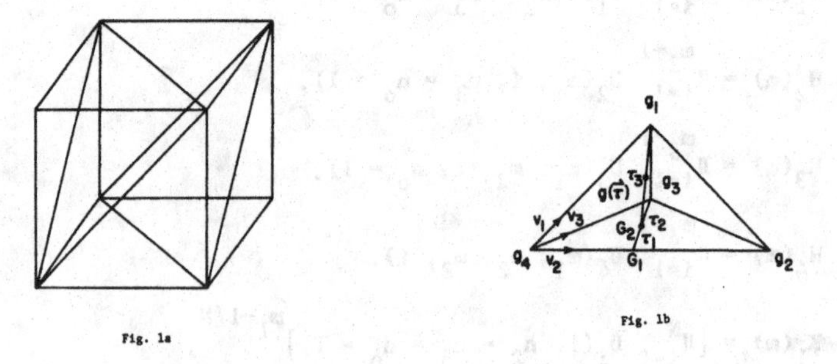

Figure 1 The division of the cube into tetrahadrons is shown in (a). A tetrahadron with vertices g_1, g_2, g_3, g_4 is depicted in (b).

$$Q = \sum_{\text{tetr.}} \frac{-1}{24\pi^2} \int_{\text{tetr.}} d^3x \, \varepsilon_{ijk} \text{tr}[w(x) \, \partial_i w^{-1}(x) w(x) \, \partial_j w^{-1}(x) w(x) \, \partial_k w^{-1}(x)]. \tag{11}$$

The direct calculation of the integrals is very lengthy even on a fast computer. But for the calculation of Q only there is a simpler method: Choose an arbitrary group element y and count how many times it occurs inside tetrahadrons. Count $r_t = +1$ if the orientation of the tetrahadron is the same in the space-time and in the group space and $r_t = -1$ otherwise. The sum over tetrahadrons $Q = \sum_t r_t$ gives the homotopy index of the configuration.

By the orientation in the group space I mean the following: Three edges of the tetrahadron starting from a vertex define three directions $\vec{v}_i$ in the group space (c.f. Fig. 1b). The sign of the volume element of these direction vectors $s = \text{sign}\left((\vec{v}_1 \times \vec{v}_2 \times)\vec{v}_3\right)$ gives the relative orientation of the tetrahadron in the group space. It is easy to see that s is independent of the choice of the vertex. The simplest way to calculate s is to multiply the vertices by g_4^{-1}. Now the three edges starting from $g_4^{-1} g_4 = 1$ are straight lines so the calculation of s is trivial. What remains to specify is how to decide whether the chosen group element y is inside or outside of the tetrahadron given by (10). Consider the tetrahadrons (y, g_2, g_3, g_4), (g_1, y, g_3, g_4), (g_1, g_2, y, g_4), (g_1, g_2, g_3, y). If all of these tetrahadrons have the same orientations in group space as the original tetrahadron (g_1, g_2, g_3, g_4), then y is inside. In such a way the decision whether $y = 1$ is inside of a tetrahadron takes three SU(2) group multiplications and five calculations of the determinant $(\vec{v}_1 \times \vec{v}_2)\vec{v}_3$ using the thermal Wilson line values taken from the lattice.

In the case of $G = SU(n)$, $n > 2$, the procedure is straightforward although lenghty. We have to decompose the thermal Wilson lines into the product of $k = n(n-1)/2$ SU(2) matrices. Then an interpolating field is defined for each SU(2) components and the topological charge is the sum of the homotopy indices of the SU(2) fields. Configurations where g^τ is not uniquely defined consists of a subset of zero measure in the configuration space.

4. DISTRIBUTION OF THE TOPOLOGICAL CHARGE ON SMALL LATTICE

4a Theoretical Considerations

To obtain the general solution of the lattice equation of motion is beyond our present capability. But even without solving lattice equations we expect that the saddle point structure differs significantly in the continuum and on the lattice. Due to the presence of the U. V. cut-off a in the lattice action, we have no zero mode corresponding to scale transformations. The way in which the breaking of the dilatation invariance destabilizes instantons is known in the two dimensional CP^{n-1} model. Calculating the action of an instanton put on a lattice it was found that the minimum of the Q = 1 topological sector is attained at the boundary of the sector [8]. Another sign of this phenomenon is that the dominant configurations of the lattice CP^{n-1} model contain pointlike gauge singularities in the continuum limit [9].

Despite this lattice peculiarity a reasonable approximation for the distribution of the topological charge can be obtained [10]. The one instanton contribution to the vacuum-to-vacuum amplitude of the continuum SU(2) gauge theory in infinite space-time in one loop order is given by [11]

$$x = \frac{\langle 0|0\rangle_1}{\langle 0|0\rangle_0} = CV \int \frac{d\rho}{\rho^5} \left(\frac{8\pi^2}{g^2}\right)^4 e^{-S/g^2 + 22/3 \ln \mu\rho} , \qquad (12a)$$

where $S = 8\cdot\pi^2$, C = 0.26 in Pauli-Villars scheme and μ is the regulator mass. In order to estimate this contribution on the lattice I make the following assumption: Introducing a mesh on a continous instanton configuration the resulting lattice field can be used in fixing the collective coordinates. I suppose that for a given set of collective coordinates there is a real saddle point of the lattice action. Accepting the existence of a saddle point the following changes have to be made in (12): i) The action of the saddle point may depend on the scale parameter $S = S(a/\rho)$. ii) The parameter C is scheme dependent. Careful two loop calculation is required to obtain its actual value. For the present I replace it by a constant to be determined later. iii) The integral over the scale parameter ρ is restricted to the interval $ax < \rho < Nay/2$ where x and y are numbers close to 1. For small N the function $S(a/\rho)$ is replaced by a constant S'. Consequently in the case of small lattice I replace (12a) by

$$x = C'V\left(\frac{S'}{g^2}\right)^4 e^{-S'/g^2} \tag{12b}$$

where C' and S' are unknown parameters.

The probabililty of finding topological charge $Q = n$ in the dilute gas approximation is given by

$$P_n = \sum_{k=0}^{\infty} q_{n+k} g_k ,$$

$$q_k = \begin{cases} C^{-1}\, x^k/k! & k \leq K \\ 0 & k > K \end{cases} , \tag{13}$$

$$C = \sum_{k=0}^{K} q_k ,$$

where the cut-off K is less than the lattice volume N^4. The topological susceptibility in this approximation is

$$\chi_t = \frac{2}{N^4} \sum_{k=0}^{K} k^2\, P_k = \frac{2X}{V} + O\left(\frac{x^{K+1}}{(K+1)!}\right) . \tag{14}$$

4b Numerical Results

In the numerical calculation I used the heat bath method described by Creutz in [2]. The explicit gauge invariant expression (9) was used to calculate the thermal Wilson lines. The probability of finding topological charge $Q = n$ on a 6^4 lattice is plotted in Fig. 2 for $n = 0,1,2$. The topological susceptibility $\chi_t = \langle Q^2\rangle/6^4$ is depicted in Fig. 3. The solid lines repesent the semiclassical results (13), (14) with $C' = .0011$, $S' = 8.1$ and $K = 20$.

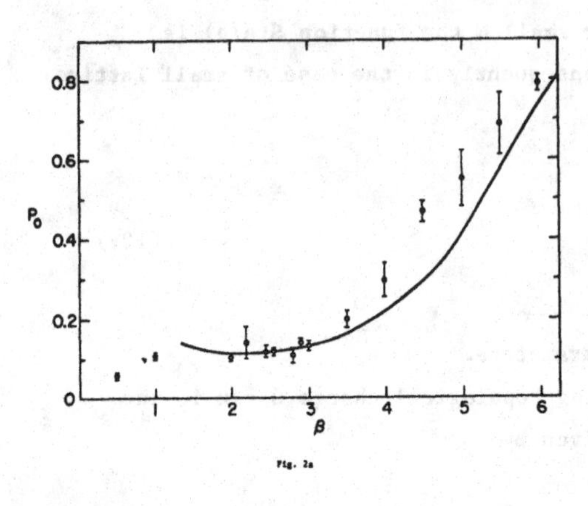

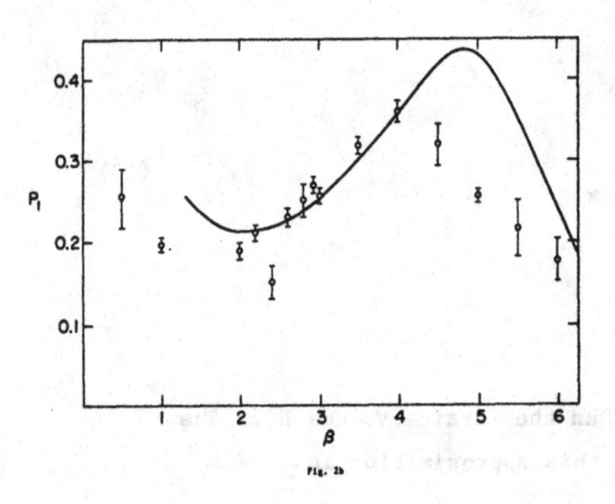

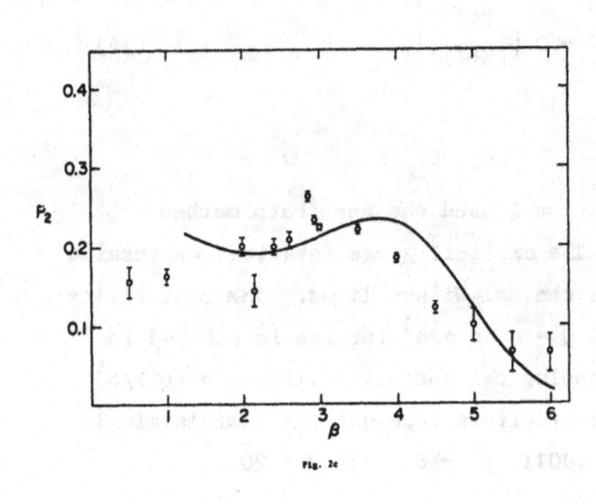

Figure 2

Probability of finding topological charge Q = n on 6^4 lattice is shown for (a) n = 0, (b) n = 1, (c) n = 2. The number of iterations was 1500 for $1. < 4/g^2 < 2.9$, and at least 3000 for $4/g^2 > 22.9$. The solid line is the result of the semiclassical expressions with C' = .0011, S' = 8.1, K = 20. The peak in the probabilities $n \neq 0$ occurs when the parameter (12b) of the truncated Poisson distribution takes the value n. The present statistical accuracy is poor to see the behaviour of Q near the critical point $4/g^2 \sim 2.25$ [11].

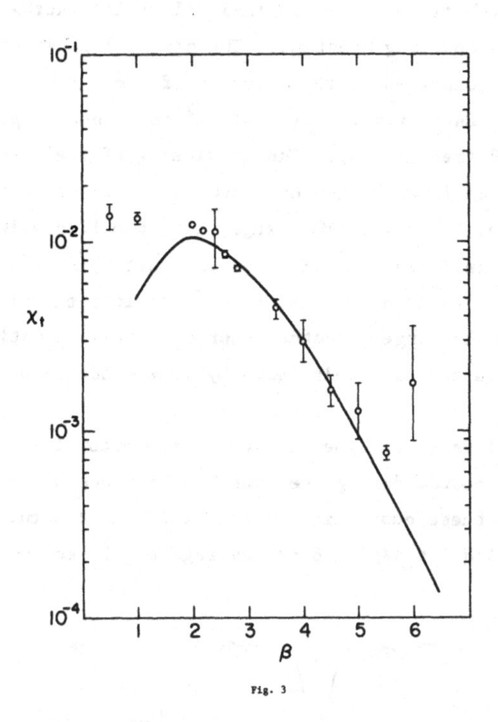

Fig. 3

Figure 3 Topological susceptibility of the 6^4 lattice. The solid line is the semiclassical result. The Monte Carlo points show exponential behavior in $4/g^2$ with slope $b = 8.1/4$ in the region $2.2 < 4/g^2 < 3.5$. The slope predicted by the renormalization group is $b = 10.77$.

The deviation from the semiclassical behaviour for $4/g^2 > 4$ comes partly from finite volume effects. Being on a finite lattice the average of the trace of the thermal Wilson line $\langle tr\Omega \rangle$ is zero for any value of the coupling constant. The critical behaviour of this quantity [12] is observed in the dynamics of the Markov process in artificial Monte Carlo time. For small g^2 one finds flip-flop behaviour in $tr\Omega$ (see Fig. 4). The critical g of the zero temperature system is where the characteristic length of the flip-flop reaches one. Whenever $tr\Omega$ changes sign the fluctation in the topological charge decreases. So I assume that this flip-flop mechanism introduces an artificial fluctation into the distribution of the topological charge. Another source of this deviation is that an extended instanton is hardly found by local updates used in the calculation.

The maximal value and the average of the acton in different Q sectors were monitored during the Monte Carlo process. I found no Q dependence in these quantities within statistical accuracy. In the region studied $1 < 4/g^2 < 6$ the average of the action

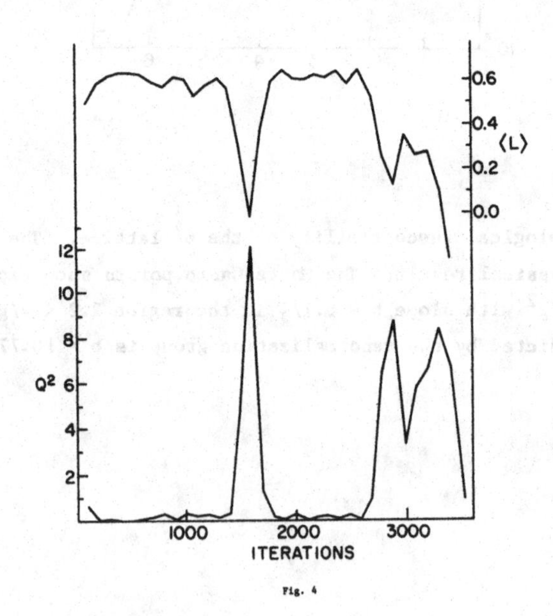

Figure 4 The averages of the thermal Wilson line and Q over 100 iterations at $4/g^2 = 6$.

$S = \sum (1-1/2\mathrm{tr}UUUU)$ is larger and the susceptibility $\chi_s = \langle S^2\rangle - \langle S\rangle^2$ is smaller than the difference of the acton between the Q = 0 and the Q = 1 sector d = 4S'. These observations suggest that P_n gives the relative volume of different Q sectors within the shell $\langle S\rangle - \chi_s < S << \langle S\rangle + \chi_s$ of the configuration space. In such a way the "classical contribution" and the "quantum fluctations" are indistinguishable in the Monte Carlo process. Another peculiarity of the Q dependence is that on calculating the average $\langle|\mathrm{tr}\Omega|\rangle$ correctly and in the Q = 0 sector only the results agree within statistical accuracy.

5. CONCLUSION

I proposed a simple procedure to calculate the topological charge of periodic lattice gauge field in four dimension. Monte Carlo calculation was carried out to determine the distribution of the topological charge on a small lattice in SU(2) gauge theory. The result was interpreted rather phenomenologically. Adjusting two relevant parameters in the continuum expression for the one instanton contribution to the path integral, the numerical results were reproduced in the interval $2.1 < 4/g^2 < 4$ on a 6^4 lattice. The renormalization of the coupling constant on such a small lattice is negligible numerically in (12). So this calculation indicates only the applicability of the dilute gas aproximation in this domain.

The measured slope of the topological susceptibility contradicts the renormalization group behaviour. This defect is traced back numerically to the smallness of the minimal action in the Q = 1 sector. But the simple renormalization group behaviour is expected only for observables without intrinsic length scale. The suggested expression for the topological charge is based on the thermal Wilson loop which feels the presence of the infrared cut-off. In other words the expression (6) gives the topological charge of the periodic system at finite temperature [13] (In fact the expression (B10) in [13] reduces to (6) in the case of periodic boundary condition in space. This boundary condition restricts the physical space of the theory excluding monopole configurations from the path integral.). In order to keep the physical size of the system unchanged one should have to adjust the lattice size too. Larger lattice size is necessary in order to push the critical value of $4/g^2$ into the weak coupling region as well. This calculation is in progress.

I thank E. Fradkin, B. Freedman and J. Kogut for useful discussions, and H. Wyld for critical reading of the manuscript. This work was supported in part by a grant from the National Science Foundation PHY82-01948.

REFERENCES

[1] K. Wilson in Recent Development in Gauge Theories edited by 't Hooft (Cargeses, 1979), Plenum Press, New York, 1980.

[2] M. Creutz Phys. Rev. D21, 2308 (1980).

[3] C. Callan, R. Dashen, D. Gross Phys. Rev. D17, 2717 (1978).

[4] M. Luscher Bern preprint BUTP-10/1981.

[5] P. Di Vecchia, K. Fabricius, G. c. Rossi, G. Veneziano, Nucl. Phys. B192, 392 (1981).

[6] E. Gava, R. Jengo, C. Omero, Nucl. Phys. B200, 107 (1982).

[7] K. Szlachanyi PhD. Th. Budapest, 1980.

[8] M. Luscher Nucl. Phys. B200[FS4], 61 (1982).

[9] J. Polonyi Urbana preprint ILL-TH-83-3.

[10] M. Evans Nucl. Phys. B208, 122 (1982).

[11] G. 't Hooft Phys. Rev. D14, 3422 (1976).

[12] E. Kovacs Phys. Lett., 125 (1982).

[13] D. Gross, R. Pisarski, L. Yaffe, Rev. Mod. Phys. 53, 43 (1981).

LOCALLY GAUGE-INVARIANT FORMULATION OF PARASTATISTICS*

O. W. Greenberg and K. I. Macrae
(talk given by O.W. Greenberg)

1. Introduction

The color degree of freedom of hadronic physics was first introduced [4] in the context of para-Fermi statistics of order three for quarks. Bose and Fermi combinations of para-Fermi quarks are in one-to-one correspondence with the color singlets of the formulation with explicit color [11]. Thus the counting of states, and the explanation of the apparent conflict with the spin-statistics theorem with quarks in the symmetric representation of SU(6) [9] are in agreement with the explicit color formulation, as is the "symmetric quark model" for baryons [4,6], which was first proposed in the context of the para-Fermi formulation. Other predictions of the two formulations which agree include the decay rate for π^0 to two photons, and, at least from the standpoint of naive counting, the ratio of the cross sections of e^+e^- to hadrons to that to $\mu^+\mu^-$. The gauge theory of color, quantum chromodynamics (QCD) differs, however, from the ungauged parastatistics formulation in predictions involving gluons, such as quark and gluon jets and the existence of glueballs. The main point of this talk is that parastatistics can be gauged and that, when gauged, it is equivalent to the corresponding Yang-Mills gauge theory, in particular, for the case of para-Fermi quarks of order three, to QCD [8].

First we give a brief review of parastatistics. Green noticed that the commutation relations between the number operators and the creation (and annihilation) operators have the same form for both the Bose and Fermi cases

$$[n_k, a_\ell^\dagger]_- = \delta_{k\ell}\, a_\ell^\dagger , \tag{1}$$

where

$$n_k = \frac{1}{2}[a_k^\dagger, a_k]_\pm + \text{const.}, \quad [A, B]_\pm = AB \pm BA, \tag{2}$$

and here and later upper (lower) signs stand for the para-Bose (para-Fermi) case. He [3] generalized (1) to allow all three indices to differ, and considered the following trilinear commutation relations

$$[[a_k^\dagger, a_\ell]_\pm, a_m]_- = -2\ \delta_{km}\, a_\ell, \tag{3}$$

$$[[a_k, a_\ell]_\pm, a_m]_- = 0. \tag{4}$$

We consider only the analog of the Fock representation of the Bose and Fermi cases, where there is a no-particle state which satisfies

$$a_k|0\rangle = 0. \tag{5}$$

Green found solutions of these commutation relations using "Green's" ansatz:

$$a_k = \Sigma_{\alpha=1}^{N}\, a_k^{(\alpha)}, \tag{6}$$

$$a_k^\dagger = \Sigma_{\alpha=1}^{N}\, a_k^{(\alpha)\dagger}, \tag{7}$$

where for equal values of α (the "Green index") the operators obey the usual commutation or anticommutation relations, but for different values of Green's index, the operators have abnormal relative commutation relations,

$$[a_k^{(\alpha)}, a_\ell^{(\alpha)\dagger}]_\mp = \delta_{k\ell}, \tag{8}$$

$$[a_k^{(\alpha)}, a_\ell^{(\beta)\dagger}]_\pm = 0,\ \alpha \neq \beta\ . \tag{9}$$

Then the expression inside the nested brackets of (3) and (4) has the form

$$\begin{aligned}[a_k^\dagger, a_\ell]_\pm &= \Sigma_\alpha \Sigma_\beta\, [a_k^{(\alpha)\dagger}, a_\ell^{(\beta)}]_\pm \\ &= \Sigma_\alpha\, [a_k^{(\alpha)\dagger}, a_\ell^{(\alpha)}]_\pm \end{aligned} \tag{10}$$

and it is clear that (3) and (4) are satisfied. The number of values over which Green's ansatz runs defines the order of the parastatistics. The order can assume any positive integer value.

The physical interpretation of parastatistics of order N is that for para-Bosons (para-Fermions) at most N particles can be in an antisymmetric (symmetric) state, while any number of particles can be in a symmetric (antisymmetric) state.

Messiah and I [5] showed that all representations in a Hilbert space of the trilinear commutation relations (3) and (4) which have a no-particle state (5) are associated with a positive integer order N, so that Green's ansatz exhausts the independent representations. As we show below, this does not mean that representations equivalent to those of Green cannot be represented in a different way.

Here is a sketch of the demonstration that Green's ansatz exhausts the independent solutions. (This demonstration makes no use of Green's ansatz.) First use (3,4,5) to show that

$$a_k a_\ell^\dagger |0\rangle = N\, \delta_{k\ell} |0\rangle, \tag{11}$$

where $N>0$ (but N is not yet shown to be an integer) and N is independent of k and ℓ. For the para-Fermi case, calculate

$$\| (a_k^\dagger)^r |0\rangle \|^2 = r!\, \frac{N!}{(N-r)!} \tag{12}$$

and for the para-Bose case,

$$\| \sum_Q \delta_Q\, a_{\mu_1}^\dagger \ldots a_{\mu_r}^\dagger |0\rangle \|^2 = (r!)^2\, \frac{N!}{(N-r)!}, \tag{13}$$

where δ_Q is the parity of the permutation Q, again using (3,4,5). These squares of norms assume negative values unless N is an integer. Thus N must be an integer.

An important application of para-Fermi fields of order three was made by showing that such fields, which contain a new three-valued degree of freedom, color, allow three quarks to be in a state which is symmetric in the other degrees of freedom, and thus resolve the spin statistics paradox which occurs in the SU(6) theory [9]. When $SU(3)_{flavor}$ and $SU(2)_{spin}$ are combined into a non-relativistic SU(6), the quarks can be assigned to the fundamental representation $\underline{6}$:

$$q_i = q_{Aa},\ A = 1,2,3 \text{ (flavor)},\ a = 1,2 \text{ (spin)}.$$

The low-lying positive parity baryons are in the symmetric three-quark representation 56 which decomposes into a spin - 1/2 octet (N,Λ,Σ,Ξ) and a spin - 3/2 decuplet (Δ, Σ^*, Ξ^*, Ω) under $SU(6) \to SU(3) \times SU(2)$. We would expect the ground state to contain quarks in S-states, so that the space wave function would be symmetric. Then the total wave function would be symmetric under permutations of the quarks, and the spin statistics theorem for spin - 1/2 quarks would be violated. Ths paradox was resolved [4] by noticing that if quarks are para-Fermions of order 3 that

$$
\begin{aligned}
B^\dagger_{ijk} &= [q^\dagger_i, [q^\dagger_j, q^\dagger_k]_+]_+ \\
&= [\sum_\gamma q_i^{(\gamma)\dagger}, \sum_{\substack{\alpha \\ \alpha\neq\beta}} \sum_\beta [q_j^{(\alpha)\dagger}, q_k^{(\beta)\dagger}]_+]_+ \\
&= 4 \sum_{\substack{\alpha\beta\gamma \\ \text{all} \\ \text{different}}} q_i^{(\gamma)\dagger} q_j^{(\alpha)\dagger} q_k^{(\beta)\dagger}
\end{aligned}
\tag{14}
$$

where the subscripts i, j, k each stand for the set of space, spin and flavor quantum numbers associated with a single quark, is a Fermi creation operator for a baryon, and is symmetric under permutations of i, j, k. The creation operator for mesons is

$$
M^\dagger_{ij} = [\bar{q}^\dagger_i, q^\dagger_j]_- = 2 \sum_\alpha \bar{q}_i^{(\alpha)\dagger} q_j^{(\alpha)\dagger}. \tag{15}
$$

Further references about parastatistics appear in [7].

There are two apparent problems which impede making Green's formulation of parastatistics into a local gauge theory: (1) neither the commutator nor the anti-commutator of the para-Fermi fields in Green's formulation has the correct number of components to couple to an SU(N) or SO(N) gauge field [1], and (2) a unitary transformation of the Green components is not again a Green component, because of the scrambling of the commutation and relative commutation relations. We pointed out that there is an exception to (1) for SO(3). (See [1].) The physically interesting case of SU(3) was gauged in the context of octonionic field theory in [10].

2. Local Gauge Theory

Usually, every index summed over occurs twice (one up, one

down) rather than once. This suggests the introduction of basis elements, so that the parafield has the form

$$\psi(x) = \sum_{\alpha} e^{\alpha} \psi_{\alpha}(x) \tag{16}$$

In later formulas we won't write the summation sign; we'll use the convention that repeated indices are summed over, but with one exception: we do not sum in $\psi^{(\alpha)}=e^{\alpha} \psi_{\alpha}(x)$. Here, we take the basis elements to be independent of space-time. Below, we discuss a formulation motivated by ideas of differential geometry in which the basis elements depend on space-time. The $\psi^{(\alpha)}$ will be Green components provided the basis elements anticommute for unequal values of their index. For the parastatistics version of SU(N) gauge theory, we choose the basis elements to form a complex Clifford algebra, with the anticommutation relations

$$[e^{\alpha}, e^{\dagger}_{\beta}]_{+} = 2\, \delta^{\alpha}_{\beta}\, I \ , \ [e^{\alpha}, e^{\beta}]_{+} = 0; \tag{17}$$

and for the SO(N) gauge theory, we choose a real Clifford algebra, with the anticommutation relations

$$[e^{\alpha}, e^{\beta}]_{+} = 2\, \delta^{\alpha\beta}\, I. \tag{18}$$

The unit elements on the right hand sides of (17) and (18) are the nxn unit matrices in the representation space of the basis elements. The representations of these Clifford algebras are given in the appendix in [8]. For the rest of this talk, we consider the SU(N) case. (The changes necessary for the SO(N) case are straightforward.) Take all the fields ψ_{α} to be Fermi fields with normal (Fermi) relative anticommutation relations. (For para-Bose fields, take all the corresponding fields to be Bose fields with normal (Bose) relative commutation relations.) Take the fields to commute with the basis elements. Now a unitary transformation of the ψ_{α} is again a Fermi field, and a unitary transformation of the basis elements is again a Clifford algebra, which removes objection (2) above. With this construction, the Green trilinear commutation relations must be altered slightly. The analog of (3) becomes

$$[[\psi^{\dagger}(x),\psi(y)]_{-} - \langle[\psi^{\dagger}(x),\psi(y)]_{-}\rangle_{o},\ \psi(z)]_{-} = -2\delta(\vec{x}-\vec{z})\psi(y) \tag{19}$$

at equal times. The symbol $\langle\ \rangle_0$ stands for the vacuum expectation value. The analog of (4), again taken at equal times, has the same form

$$[[\psi(x),\ \psi(y)]_-,\ \psi(z)]_- = 0. \tag{20}$$

To generate the gauge interaction, we follow the procedure of Yang and Mills, and consider space-time dependent gauge transformations $U(x)\ \psi(x)\ U^\dagger(x)$, where

$$\begin{aligned} U(x) &= \exp\{i \frac{1}{4n} \operatorname{tr} \int [\psi^\dagger(x),[t(x),\psi(x)]_-]_- d^3x\} \\ &= \exp\{i \int \psi^{\alpha\dagger}(x)\ t_\alpha{}^\beta(x)\ \psi_\beta(x)\ d^3x\}, \end{aligned} \tag{21}$$

and

$$t(x) = e^\alpha\ t_\alpha{}^\beta(x)\ e^\dagger_\beta,\ t_\alpha{}^\beta = t_\beta{}^{\alpha *},\ t_\alpha{}^\alpha = 0. \tag{22}$$

We find

$$U(x)\ \psi(x)\ U(x)^\dagger = \psi^U(x) = e^\alpha\ (U\psi)_\alpha(x)\ , \tag{23}$$

where

$$(U\psi)_\alpha(x) = U_\alpha{}^\beta(x)\ \psi_\beta(x),\ \text{and}\ U_\alpha{}^\beta(x) = (\exp\ i\ t)_\alpha{}^\beta(x). \tag{24}$$

The kinetic term in the Lagrangian for the Dirac field ψ is $\frac{1}{2}(i[\bar\psi,\not\partial\psi]_- - i\langle[\bar\psi,\not\partial\psi]_-\rangle_o)$. Under a gauge transformation the kinetic term acquires the following non-invariant term due to the space-time dependence of U,

$$i :\bar\psi^\beta\ (U^\dagger)_\beta{}^\alpha\ (\not\partial U_\alpha{}^\delta)\ \psi_\delta:.$$

This non-invariant term can be canceled, as usual, by the introduction of the gauge field associated with changing the derivative to the gauge-covariant derivative

$$i\ \partial_\mu \delta_\alpha{}^\beta \rightarrow i\ \partial_\mu \delta_\alpha{}^\beta - g\ A_{\mu\alpha}{}^\beta\ ,\text{where}\ A_{\mu\alpha}{}^\beta = \frac{\lambda_\alpha^{a\beta}}{2} A_\mu^a\ , \tag{25}$$

and requiring $A_{\mu\alpha}{}^{\beta}$ to transform as

$$A_{\mu\alpha}{}^{\beta} \rightarrow U_{\alpha}{}^{\gamma} A_{\mu\gamma}{}^{\delta} (U^{\dagger})_{\delta}{}^{\beta} + \frac{1}{ig} (\partial_{\mu} U)_{\alpha}{}^{\gamma} U_{\gamma}{}^{\beta}. \tag{26}$$

The trilinear commutation relations for the gauge field are,

$$[\mathrm{tr}\ \frac{2}{n} [\dot{A}_{\mu}(x), A_{\nu}(y)]_{+} - \langle\mathrm{same}\rangle_{o}, A_{\lambda}(z)] ,$$

$$= 2ig_{\mu\lambda} \delta^{3}(\vec{x}-\vec{z}) A_{\nu}(y), \tag{27}$$

at equal times; the analogous trilinear commutators with only fields or with only time derivatives of fields vanish. What remains to be shown is that the interaction term can be written in terms of parafields. Let

$$A_{\mu} = \tfrac{1}{2} e^{\alpha} A_{\mu\alpha}{}^{\beta} e_{\beta}^{\dagger},\ A_{\mu\alpha}{}^{\beta\dagger} = A_{\mu\beta}{}^{\alpha},\ A_{\mu\alpha}{}^{\alpha} = 0 . \tag{28}$$

Then the interaction term is

$$\frac{1}{4n}\ \mathrm{tr}\ ([\bar{\psi},[A_{\mu},\psi]_{-}]_{-} - \langle\mathrm{same}\rangle_{o})$$

$$= :\bar{\psi}^{\alpha} A_{\mu\alpha}{}^{\beta} \psi_{\beta}: \tag{29}$$

The pure gauge term in the Lagrangian is

$$-\frac{1}{4n}\ \mathrm{tr}\ G_{\mu\nu} G^{\mu\nu}, \tag{30}$$

where

$$G_{\mu\nu} = \partial_{\mu}A_{\nu} - \partial_{\nu}A_{\mu} + ig[A_{\mu},A_{\nu}]_{-}$$

$$= \tfrac{1}{2} e^{\alpha} [\partial_{\mu}A_{\nu\alpha}{}^{\beta} - \partial_{\nu}A_{\mu\alpha}{}^{\beta} + ig(A_{\mu\alpha}{}^{\delta}A_{\nu\delta}{}^{\beta} - A_{\nu\alpha}{}^{\delta}A_{\mu\delta}{}^{\beta})]e_{\beta}^{\dagger}$$

$$= \tfrac{1}{2} e^{\alpha} (G_{\mu\nu})_{\alpha}{}^{\beta} e_{\beta}^{\dagger}. \tag{31}$$

We have now shown that the entire Lagrangian of an SU(N) Yang-Mills field interacting with a spinor field can be expressed in terms of parafields, as follows:

$$\mathcal{L} = \frac{1}{2n}\ \mathrm{tr}\ [\bar{\psi}, i\hat{D}\psi]_{-} - \frac{1}{4n}\ \mathrm{tr}\ G_{\mu\nu} G^{\mu\nu}, \tag{32}$$

where

$$\hat{D}_\mu = \tfrac{1}{2} e^\alpha \, (\delta_\alpha{}^\beta \, \partial_\mu + i \, g \, A_{\mu\alpha}{}^\beta) \, e^\dagger_\beta. \tag{33}$$

A similar construction can be given if scalar fields are also present.

Parastatistics can also be gauged in the functional integral approach: Replace the fermionic coefficients ψ_α by standard Grassmann elements, and the bosons by standard c-numbers. This implicitly defines a para-Grassmann algebra and a para-scalar algebra. The brackets of these algebras are the same as those in (8) and (9), except that the $\delta_{k\ell}$ should be replaced by zero.

There is a way in which the present parafield description of Yang-Mills theory differs from the usual description. Since the basis elements are nilpotent, the (N+1)st power of the spinor field vanishes identically. This problem can be avoided as far as SU(N)-singlet states are concerned as follows: introduce operators for the "baryons" and "mesons"

$$b(x_1, x_2, \ldots x_N) = [\psi(x_N),[\psi(x_{N-1}), [\ldots[\psi(x_2),\psi(x_1)]_+\ldots]_+]_+]_+$$

$$=[e^{\alpha_N}, [\ldots[e^{\alpha_2}, e^{\alpha_1}]_-]_+\ldots]_{(-)^{N+1}} \, \psi_{\alpha_N}(x_N)\ldots\psi_{\alpha_2}(x_2) \, \psi_{\alpha_1}(x_1)$$

$$= 2^{N-1} e^{\alpha_N} \ldots e^{\alpha_2} e^{\alpha_1} \, \varepsilon^{\alpha_N\ldots\alpha_2\alpha_1} \, \psi_{\alpha_N}(x_N)\ldots\psi_{\alpha_2}(x_2) \, \psi_{\alpha_1}(x_1), \tag{34}$$

where the $(-)^{N+1}$ in the second line indicates that the last bracket is a commutator for N even and an anticommutator for N odd. Let

$$E = e^1 \, e^2 \, \ldots \, e^N \tag{35}$$

be the element analogous to γ_5 in the usual Dirac algebra. Then

$$B(x_1, x_2, \ldots x_N) = \frac{1}{2^{N-1}} \mathrm{tr}_n \, (Eb(x_1, x_2, \ldots x_N))$$

$$= \varepsilon^{\alpha_N \ldots \alpha_2 \, \alpha_1} \, \psi_{\alpha_N}(x_N) \ldots\psi_{\alpha_2}(x_2) \, \psi_{\alpha_1}(x_1), \tag{36}$$

where the ε tensor is antisymmetric and $\varepsilon_{N,\ldots 2,1} = 1$, is a field for a color-singlet baryon, and is free of basis elements. For mesons, the analogous construction is simpler. The operator

$$M(y, x) = \frac{1}{2n} \text{tr} \, [\bar{\psi}(y), \psi(x)]_- = :\bar{\psi}^\alpha(y) \, \psi_\alpha(x): \tag{37}$$

serves as a field for a color-singlet meson. (To make the baryon and meson operators gauge invariant, the usual path-ordered exponential factors must be supplied.)

From the point of view of differential geometry, the elements are a basis for the SU(N) space at each point, and, in general, will be space-time dependent. Then the gauge (or connection) field is the matrix which gives the change of the basis elements in a given direction. In general the basis elements at finitely separated points are related in a way which depends on the path taken in going between the two points. This is expressed by writing

$$de^\alpha(x) = ig \, e^\beta(x) \, A_{\mu\beta}{}^\alpha(x) \, dx^\mu. \tag{38}$$

Then, the gradient of the spinor field $\psi = e^\alpha \psi_\alpha$ contains two terms. One is the gauge (or connection) field and the other is the change of the components ψ^α,

$$d\psi = d(e^\alpha \psi_\alpha) = e^\alpha \, (\partial_\mu \psi_\alpha) \, dx^\mu + ig \, e^\beta A_{\mu\beta}{}^\alpha \, \psi_\alpha \, dx^\mu$$

$$= e^\alpha \, (D_\mu \psi)_\alpha \, dx^\mu, \tag{39}$$

where $(D_\mu \psi)_\alpha$ is the usual gauge-covariant derivative. This motivates the notation $d_\mu \psi = e^\alpha \, (D_\mu \psi)_\alpha$. The covariantly coupled kinetic term for the spinor field is

$$\frac{1}{2n} \text{tr} \, [\bar{\psi} \, , \not{d} \psi]_-. \tag{40}$$

In a similar way, the commutator of the d_μ's generates the gauge field tensor (or curvature),

$$[d_\mu, d_\nu]_- \, e^\alpha = i \, g \, e^\beta \, (G_{\mu\nu})_\beta{}^\alpha \tag{41}$$

and the gauge field term in the Lagrangian is

$$- \frac{1}{4ng^2} \text{tr} \, |[d_\mu, d_\nu]_- \, e^\alpha|^2. \tag{42}$$

This completes the demonstration that parastatistics can be gauged to give SU(N) gauge theory.

Our conclusion that parastatistics can be gauged differs from that of Govorkov [2].

References

[1] P. G. O. Freund, Phys. Rev. D 13, 2322 (1976).

[2] A. B. Govorkov, Dubua preprints P2-81-749, P2-82-296, and E-2-82-470.

[3] H. S. Green, Phys. Rev. 90, 270 (1953).

[4] O. W. Greenberg, Phys. Rev. Lett. 13, 598 (1964).

[5] O. W. Greenberg and A. M. L. Messiah, Phys. Rev. 138, B1155 (1965).

[6] O. W. Greenberg and M. Resnikoff, Phys. Rev. 163, 1844 (1967).

[7] O. W. Greenberg and C. A. Nelson, Phys. Reports 32C, 69 (1977).

[8] O. W. Greenberg and K. I. Macrae, Nucl. Phys. B219, 358 (1983).

[9] F. Gürsey and L. A. Radicati, Phys. Rev. Lett. 13, 173 (1964).

[10] M. Günaydin and F. Gürsey, Phys. Rev. D9, 3387 (1974); F. Gürsey , in Proceedings of the Johns Hopkins Workshop on Current Problems in High Energy Particle Theory, 1974, edited by G. Domokos and S. Kövesi-Domokos (Johns Hopkins Univ. Press, Baltimore, 1974), p. 15.

[11] M. Y. Han and Y. Nambu, Phys. Rev. 139, B1006 (1965).

* This work was supported in part by the National Science Foundation.

Center for Theoretical Physics,
Department of Physics and Astronomy,
University of Maryland,
College Park, Maryland 20742.

AFFINE ALGEBRAS and STRONG INTERACTION THEORIES

L. Dolan
The Rockefeller University, New York, N.Y. 10021

Abstract

A new infinite parameter symmetry, the Kac-Moody Lie algebras, appears in several different connections with theories of the strong interactions. Its explicit representation in terms of the dual string is reviewed here in simple language. A nonlocal nonlinear realization of a subalgebra of the affine algebras on the self-dual class of solutions of the Yang Mills theory is also reviewed. It is remarked that this structure occurs naturally in a Kaluza-Klein search for the new symmetry on the full gauge theory. The existence of the same invariance in these different models of the hadrons may serve to unify the descriptions and lead to solvability in the nonperturbative regime.

1. Introduction

Although we believe gauge theories play a fundamental role in hadronic physics, there is no systematic controlled approximation in which to calculate most of the current experimentally accessible phenomena. At distances on the order of the size of elementary particles, the strong interactions, distinguished by their large coupling, require non-perturbative analysis. The invariance properties of a model provide a guide to the nature of the exact (non-perturbative) solution; thus any new continuous symmetry in four dimensional Yang Mills theory would be extremely valuable information. Presumably if gluons confine quarks non-perturbatively in QCD, they will also confine themselves in a pure glue theory with no quarks; and any symmetry responsible for this confinement will be manifest already in the pure gauge theory.

There is a candidate for such a symmetry and so far it has been realized explicitly on the self-dual solutions in 4-dimensional Euclidean space ($F_{\mu\nu} = 1/2\, \varepsilon_{\mu\nu\alpha\beta} F^{\alpha\beta} \equiv \tilde{F}_{\mu\nu}$). The generators of these infinitesimal transformations form a (parabolic) subalgebra of an infinite parameter affine algebra known to mathematicians as a Kac-Moody Lie algebra: $G \otimes \mathbb{C}[t,t^{-1}] \oplus \mathbb{C}_z$, where G is a finite parameter semi-simple algebra which for Yang-Mills is SU(N). For C_{abc} the structure constants of G, the generators M^n_a of the affine algebra obey

$$[M_a^n, M_b^m] = C_{abc} M_c^{n+m} + n\,\delta_{n,-m}\,\delta_{ab}\,P \qquad (1.1a)$$

$$[P, M_a^n] = 0 \qquad (1.1b)$$

where n,m = $-\infty$, ...-1,0,1,...∞. What occurs naturally as a symmetry of the self-dual solutions is the subalgebra given by (1.1) when n,m = 0,1,..∞. That is to say there exists an infinite set of transformations $A_\mu \to A'_\mu \simeq A'_\mu + \Delta^n_a A_\mu$ such that if A_μ is a solution to $F_{\mu\nu} = \tilde{F}_{\mu\nu}$ so is A'_μ and the generators $M^n{}_a \equiv -\int d^4x \Delta^n{}_a A_\mu(x) \frac{\delta}{\delta A_\mu(x)}$ close "half" an affine algebra (modulo gauge transformations)[1]. The first transformation is given here as an example. In complex coordinates, $\sqrt{2}y = x_4 - ix_3$, $\sqrt{2}\,z = x_2 - ix_1$, and $\Delta'_a A_y = D_y\Omega_a$, $\Delta'_a A_{\bar{y}} = -D_{\bar{y}}\Omega_a$, $\Delta'_a A_z = -D_z\Omega_a$, $\Delta'_a A_{\bar{z}} = D_{\bar{z}}\Omega_a$ where $A_\mu = A^a_\mu T^a$, $T^a = \frac{\sigma}{2i}$ for SU(2). $\Omega_a = \Omega^+_a$ is given in terms of Yang's D and $\bar{D}$ functions:

$$D(x) = P e^{\int^x_{\uparrow} \{d\bar{z}' A_{\bar{z}} + dy' A_y\}} \quad ; \quad \bar{D} = (D^+)^{-1}, \qquad (1.2)$$

path has fixed $\bar{y}$,z

$$\Omega_a = \tfrac{1}{2}\{\bar{D}^{-1}\Lambda_a\bar{D} + D^{-1}\Lambda^+_a D\}$$

$$\partial_z \Lambda_a = [J^{-1}\partial_y J, T_a] \quad ; \quad J = D\bar{D}^{-1}. \qquad (1.3)$$

Clearly, for this symmetry to be useful in constructing a non-perturbative solution to QCD, we must first be able to extend it off the self-dual set to the full Yang Mills theory. A program to carry this out using the general coordinate invariance of multi-dimensional gravity in Kaluza-Klein theories to find a new invariance of non-abelian gauge theory is underway. This has been reported on elsewhere[2], but we are heartened to mention here that the complicated non-linear non-local structure which appears in (1.2) arises naturally in the solution of the characteristic curves associated with partial differential equations for $\Delta A_\mu{}^a$ generated by the Kaluza-Klein approach.

In this talk, I will describe in simple language the appearance of this same Kac-Moody symmetry algebra in connection with another model for hadrons, the dual string. It was in this context that the mathematicians[3,4] first wrote down representations of (1.1). In Section 2,I give a pedestrian explanation of their results which may be more transparent for physicists. It will also serve to motivate why the infinite parameter invariance should be present in the full non-abelian gauge theory and may unify these two different descriptions of the strong interactions.

2. The Dual String

In the late 1960's, the dual resonance model was invented, without reference to non-abelian gauge theory, to describe hadronic scattering. The N-point functions for N scalar particles are given in terms of a vertex operator. $V_0(k,z)$. It is this operator whose coefficients when expanded in a power series in z form a representation of a Kac-Moody algebra. General S-matrix elements are calculated from the N-point functions using a factorization condition. They provide an approximation in which all resonances are infinitely narrow and the Regge trojectories (the square mass versus spin plot) are linear.

The spectrum of the dual model consists of an infinite number of states which lie in a Fock space generated by an infinite number of creation operators $a^{+\mu}_m$. The vertex operator is given in terms of the creation and the annihilation operators a^μ_m . Thus their commutation relations will determine the affine algebra of the coefficients. The commutation relations of a and a^+ can be given in the compact form, for $m,n = 0, \pm 1 \ldots \pm \infty$,

$$[\alpha^\mu_m , \alpha^\nu_n] = m g^{\mu\nu} \delta_{m,-n} \tag{2.1}$$

Here $\alpha^\mu_m \equiv \sqrt{m}\, a^\mu_m$ and $\alpha^\mu_{-m} = \sqrt{m}\, a^{+\mu}_m$ for $m = 1,2,\ldots\infty$. The momentum operator is defined by $\alpha^\mu_0 = \sqrt{2\alpha'}\, p^\mu$ and the position operator has $[x^\mu , p^\nu] = i g^{\mu\nu}$

The vertex operator is then given by

$$V_0(k,z) = e^{\sqrt{2\alpha'}\, k_\mu \sum_{n=1}^{\infty} \alpha^\mu_{-n} z^n} e^{ik\cdot x} z^{2\alpha' k\cdot p} e^{-\sqrt{2\alpha'}\, k_\mu \sum_{n=1}^{\infty} \frac{\alpha^\mu_n}{n} z^{-n}} . \tag{2.2}$$

For the string in one dimension ($\mu = 1$) and conformal spin $\alpha' k^2 = 1$ $V_0(k,z)$ can be written as

$$V_0(\gamma, z) = e^{\sum_{n=1}^{\infty} \frac{z^n}{n}\gamma(-n)} \; e^{[\gamma + \ln z \frac{\partial}{\partial \gamma}]} \; e^{-\sum_{n=1}^{\infty} \frac{z^{-n}}{n}\gamma(n)} \tag{2.3}$$

Here,

$$\gamma(n) = \sqrt{2\alpha'}\, k \cdot \alpha_n \;, \quad \gamma = i k \cdot x \;, \quad \frac{\partial}{\partial \gamma} = k \cdot p \; 2\alpha' \tag{2.4}$$

Define coefficients $X_n(\pm)$ from

$$V_0(\gamma, z) = \sum_{n=-\infty}^{\infty} z^{-n} X_n(+)$$

and

$$V_0^{\dagger}(\gamma, z) = \sum_{n=-\infty}^{\infty} z^{n} X_n(-) \, . \tag{2.5}$$

These coefficients together with $\gamma(n)$ form a representation of the affine algebra. From (2.1) - (2.5), then $[\gamma, \frac{\partial}{\partial \gamma}] = -2$, and the Kac-Moody Lie algebra $SL(2,C) \otimes \mathbb{C}[t,t^{-1}] \oplus \mathbb{C}_z = \hat{S}L(2,C)$ appears as

$$[\gamma(n), \gamma(m)] = 2n\,\delta_{n,-m}\, I \tag{2.6a}$$

$$[X_n(+), X_m(-)] = \gamma(n+m) + n\,\delta_{n,-m}\, I \tag{2.6b}$$

$$[\gamma(n), X_m(\pm)] = \pm 2 X_{n+m}(\pm) \tag{2.6c}$$

and

$$[X_n(+), X_m(+)] = 0$$

$$[X_n(-), X_m(-)] = 0 \, . \tag{2.6d}$$

A basis for $SL(2,R)$ or $SL(2,C)$ is

$$e = \begin{pmatrix} 0 & 1 \\ 0 & 0 \end{pmatrix}, \quad f = \begin{pmatrix} 0 & 0 \\ 1 & 0 \end{pmatrix}, \quad h = \begin{pmatrix} 1 & 0 \\ 0 & -1 \end{pmatrix}.$$

Then $[h,e] = 2e$, $[h,f] = -2f$, $[e,f] = h$, and $\mathrm{Tr}\, ef = 1$, $\mathrm{Tr} h^2 = 2$.

For $X(0)$, $Y(0)$ generators of $SL(2,C)$, the commutations for $\hat{S}L(2,C)$ are, from (1.1),

$$[X(m), Y(n)] = [X(0), Y(0)] \otimes t^{n+m} + m\,\delta_{n,-m}\, I \;\mathrm{Tr}\, X(0) Y(0) \, . \tag{2.7}$$

Therefore, from (2.6), $\gamma(n)$, $X\,(+)$ and $X\,(-1)$ are generators of $SL(2,c)$ where $(0) = h$, $X_0(+) = e$ and $X_0(-) = f$.

In the QCD picture of hadrons, the quarks are held together "inside" the hadron by the gluon color flux. These lines of color force do not spread out in space as in an electric dipole but rather are confined in a narrow tube or thick string. In the $N \to \infty$ limit of an SU(N) non-abelian color group, the tube becomes infinitely thin and looks like a real string. But the connection between the dual model and this field theory limit is not precisely understood.

The explicit realization of the affine algebra in the full gauge theory may elucidate this tie. And the existence of this algebra in the string makes the $F_{\mu\nu} = \tilde{F}_{\mu\nu}$ restriction in the gauge theory seem not to be fundamental.

3. Nonperturbative Solvability

An infinite parameter invariance is a hallmark of exact solvability. In the two-dimensional non-linear sigma model the loop algebra is present and leads to restrictions on the S-matrix such that it can be solved completely. In the spin models of statistical physics: the Ising model and the Baxter models, an infinite set of conserved charges signals the existence of action-angle variables which permits the diagonalization of the Hamiltonian. The Kramers-Wannier self-duality of these models is thought to be shared by the four-dimensional. SU(N) theory and a connection can be established between this self-duality and the existence of an infinite dimensional symmetry.

It thus seems reasonable that affine algebras will be relevant for a nonperturbative solution of the strong interactions. That they probably exist in the gauge theory is supported by the above arguments.. Their usefulness in finding a non perturbative solution is signaled by the examples of section 3 and much recent work on the Toda chain and other simple systems where the Kac-Moody algebras are central to the construction of the Lax pair.

Acknowledgements

I would like to acknowledge conversations with J. Lepowsky.

Work supported in part by the Department of Energy under Contract Grant No. DE-AC02-83ER40033.B000.

References

[1] L. Dolan, "Kac-Moody Algebras and Exact Solvability in Hadronic Physics", to be published in Physics Reports; and references therein.

[2] L. Dolan, "Kaluza-Klein Theories as a Tool to Find New Gauge Theory Symmetries", to be published in the Proceedings of Orbis Scientiae 1983, Coral Gables.

[3] I.B. Frenkel and V.G. Kac, Inv. Math. 62, 23 (1980);
J. Lepowsky and R.L. Wilson, Com. Math. Phys. 62, 43 (1978).

[4] For more expository papers, see also I.B. Frenkel, "Representations of Kac-Moody algebras and Dual Resonance Models", Princeton preprint and J. Lepowsky, "Some Constructions of the affine Lie algebra $A_1^{(1)}$", Rutgers preprint.

GLUON CONDENSATION IN QCD AND SUPERSYMMETRIC QCD

Yoichi Kazama
Department of Physics, Kyoto University, Kyoto 606, Japan

§1 Introduction

Although non-perturbative study of field theories has a long history, it is a feeling shared by many people, perhaps by all of us at this workshop, that it is becoming increasingly more critical for the deep understanding of physical laws of nature. It is probably not an exaggeration to say that it is an emblem of the particle physics of the 80's. This I believe is an inevitable trend. The development of non-abelian gauge theories in the 70's has brought us to the stage where we cannot help talking about the derivation of the spectrum of the world with naturalness and economy (unification). This is outside the realm of the perturbation theory, in which one particle requires one field with prescribed characteristics. More dynamical aspects of gauge theories must be understood, and QCD and supersymmetric versions of it will certainly play primary roles as prototypical theories.

Ther are of course many different attack routes as have been and will be discussed in this workshop. What I would like to discuss is one of such attmepts, based on the study of the gluon condensation, $\langle 0|(F^a_{\mu\nu})^2|0\rangle$, which is one of the important "order parameters" of QCD and super QCD. I will start by giving some theoretical understanding of gluon condensation in ordinary QCD, then go on to the main part of this talk, which will be on super QCD. This study, however, in the end, will lead back to suggest an unconventional possibility in ordinary QCD.

§2. Role of Gluon Condensation and a Theoretical Understanding in QCD

Let us begin by giving some theoretical understanding of gluon condensation in ordinary QCD and what that implies. The gluon content of the vacuum can be probed in many ways. Since the gluons do not carry any flavor, any gauge invariant operator, local or non-local, will do. Among them, perhaps the simplest one is the operator $(F^a_{\mu\nu})^2$ and let us see if this has non-vanishing expectation value in the vacuum. (In the following, when we write $\langle 0|(F^a_{\mu\nu})^2|0\rangle$, it really means the normal product $\langle 0|N((F^a_{\mu\nu})^2)|0\rangle$, which is defined to vanish in perturbation theory.) As we all know, this quantity has been measured experimentally through the QCD sum rule developed by Shifman, Vainstein, and Zakharov [1]. Fitting the sum rule to the e^+e^- data around the charmonium region, these authors obtained

$$\langle 0|\frac{\alpha_s}{\pi}F_{\mu\nu}^2|0\rangle = \langle 0|\frac{2\alpha_s}{\pi}(H^2-E^2)|0\rangle \simeq 0.012\,\mathrm{Gev}^4 > 0\,, \qquad (2.1)$$

i.e., it is non-zero and positive, or "magnetic".

Theoretically many authors have discussed this magnetic condensation by various means [2]. I simply do not have time to review them all. So allow me to breifly review the method developed by R. Fukuda and myself a few years ago [3] based on the trace anomaly equation, because I will need it later for supersymmetric QCD.

First let us see that the above experimental result is theoretically reasonable as well. The well-known trace anomaly equation in the vacuum is of the form

$$\langle 0|\theta^\mu_\mu|0\rangle = \frac{2\beta(g)}{g}\langle 0|\tfrac{1}{4}F_{\mu\nu}^2|0\rangle. \qquad (2.2)$$

Since the vacuum is Poincaré invariant, we get the vacuum energy density by dividing by four;

$$\varepsilon_{vac} = \langle 0|\theta_{00}|0\rangle = \frac{\beta(g)}{2g}\langle 0|\tfrac{1}{4}F_{\mu\nu}^2|0\rangle\,. \qquad (2.3)$$

Now the fact that $\beta(g)<0$ tells us that the energy density goes down below zero (which is the perturbative value) if $\langle 0|F_{\mu\nu}^2|0\rangle>0$, agreeing with the experimental result. But this does not mean that it actually happens. Anomaly equation by itself is perfectly consistent with both sides vanishing.

To see that it does happen, we might try to use the standard method. I.e., we introduce a source J for $F_{\mu\nu}^2$, compute the generating functional

$W[J]$, and Legendre-transform it to get the effective potential for $F^2_{\mu\nu}$. This program is quite difficult to carry through. Instead we shall use a trick.

The trick is to derive the trace anomaly equation in the presence of the space-time independent source J_0 for $F^2_{\mu\nu}$. Once we do this, just as before, we will get the vacuum energy density, but now in the presence of the source, and by Legendre-transforming it, we get the effective potential, or rather, as it turns out, a differential equation for it. Starting from the generating functional

$$Z[J] = \exp iW[J] = \int \mathcal{D}A^\mu_0 \, \delta(n\cdot A_0) \exp\left(-i(1+J_0)\int d^4x \tfrac{1}{4}F^2_{\mu\nu}\right) \times \exp i\int j_{\mu 0} A^\mu_0 \, d^4x \quad , \tag{2.4}$$

we can more or less follow the procedure developed by Collins, Duncan and Joglekar [4], for the trace anomaly without the source term. The only thing to be careful about here is that we must properly take care of the renormalization of multiple insertions of $\int d^4x F^2_{\mu\nu}$ [3]. We simply quote the result:

$$\begin{aligned} \mathcal{E}(J) &= \langle 0|\theta_{00J}|0\rangle = -\frac{2\beta(g_J)}{g_J}(1+J)F \\ F &\equiv \langle 0|-\tfrac{1}{4}F^2_{\mu\nu}|0\rangle \quad , \quad g_J^2 = \frac{g^2}{1+J} \quad , \end{aligned} \tag{2.5}$$

where g and J are the coupling constant and the source renormalized at scale μ. By Legendre transformation, we get

$$V(F) = \mathcal{E}(J) + JF = -\frac{\beta(g_J)}{2g_J}(1+J)F + JF \quad , \tag{2.6}$$

with $J=dV/dF$. This form does not change in the presence of massless quarks. They manifest themselves only though the β function. Now the point is that Eq.(2.6) is recognized as a non-linear differential equation for V(F). Expanding the β function $\beta(g_J)=b_0g_J^3+b_1g_J^5+\ldots$, it takes the form

$$V(F) = -\frac{1}{2}\left(b_0 g^2 + \frac{b_1 g^4}{1+\frac{dV}{dF}} + \cdots\right)F + F\frac{dV}{dF} \quad . \tag{2.7}$$

By taking μ to be large, $g^2(\mu)$ can be made small and we may look for a solution in the region where $F \leq 0$ (magnetic condensation). We get

$$V(F) = -F + \frac{1}{2} b_0 g^2 F \left(\ln \frac{-g^2 F}{\mu^4} + C \right) , \tag{2.8}$$

with the stationary point

$$F = F_s = -\mu^4 \frac{1}{g^2} \exp\left(\frac{2}{b_0 g^2} - C - 1 \right) < 0 . \tag{2.9}$$

This potential is sketched in Fig. 1. We see that there is a stable unique non-perturbative vacuum with $F_s < 0$. Alghouth we cannot get the value of F_s as we have truncated $\beta(g)$, for large enough μ the correction for the part multiplied by μ^4 is small and the qualitative feature does not change.

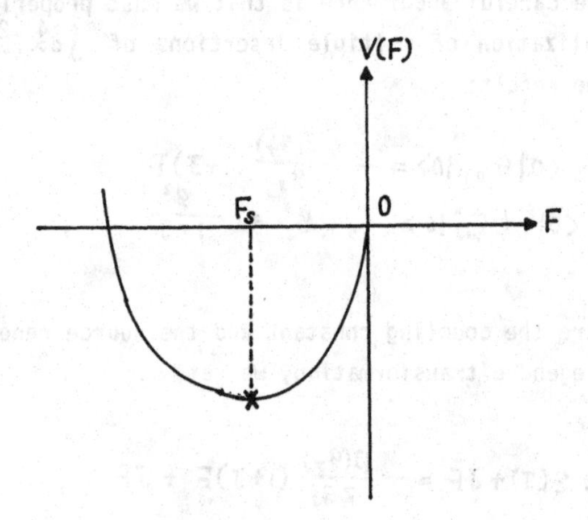

Fig. 1. Effective potential for $F = \langle 0 | -\frac{1}{4} F_{\mu\nu}^2 | 0 \rangle$.

Now accepting that the color singlet gluon condensation occurs, let us ask what it implies. For this purpose, let me briefly describe the recent work of R. Fukuda [5], which, although some assumptions are made, gives us a very clear picture of what is going on.

Let us take an axial gauge $A_3^a = 0$, in which the relation between the field strength and the vector potential is very simple; $F_{3\alpha}^a = \partial_3 A_\alpha^a$

(α=0,1,2). What Fukuda assumes is that the space-time independent (i.e. zero momentum) mode $A_\mu^{(0)}$ dominates in the gluon condensation. This is a reasonable assumption since the gluon condensation must be a low energy phenomenon. So let us split the potential

$$A_\mu^a(x) = A_\mu^{(0)a} + A_\mu'^a(x). \tag{2.10}$$

What condenses is the globally color singlet "radial" part of $A_\alpha^{(0)a}$. Alternatively one may write

$$\langle 0|\, A_i^{(0)a} A_j^{(0)b} |0\rangle = \delta^{ab}\delta_{ij} D^{(0)} \neq 0 \ . \tag{2.11}$$

Then he observes, by using canonical equal time commutation relations, that for small p the following equation holds:

$$[\dot{F}_{3i}^a(x), F_{3j}^b(0)]_{x_0=0} \underset{p\to 0}{\cong} \frac{1}{i}\delta^{ab}\delta_{ij} C_2(G) g^2 D^{(0)} \delta^3(x). \tag{2.12}$$

In other words, when the zero momentum mode condenses, the magnetic field strength is a good canonical coordinate in the low momentum region.

To make a full use of this observation, he utilizes the field srength formulation of Yang-Mills theory developed by Halpern [6]. The procedure is as follows: Start from the first order formalism for the path integral representation of the generating functional

$$I = \int \mathcal{D}A^{(0)} \mathcal{D}A' \mathcal{D}F_{\mu\nu} \exp(-\tfrac{i}{4}\int d^4x F_{\mu\nu}^2)\, \delta(F_{\mu\nu} - F_{\mu\nu}(A))\, \delta(A_3). \tag{2.13}$$

Integrating out A'^a, one gets

$$I = \int \mathcal{D}A^{(0)} \mathcal{D}F_{\mu\nu}\, \delta(D^\mu(A^{(0)})\tilde{F}_{\mu\nu}) \exp(-\tfrac{i}{4}\int d^4x F_{\mu\nu}^2). \tag{2.14}$$

The Bianchi-like δ function constraint is then solved by the introduction of the dual vector potential B_μ^a

$$I = \int \mathcal{D}A^{(0)} \mathcal{D}F_{\mu\nu} \mathcal{D}B_\mu \exp\left(i \int d^4x\, \overline{\mathcal{L}}(F_{\mu\nu}, B_\mu)\right), \qquad (2.15)$$
$$\overline{\mathcal{L}} = -\frac{1}{4} F_{\mu\nu}^2 + D^{\mu ab}(A^{(0)}) B^{b\alpha} \tilde{F}^a_{\mu\alpha} \quad .$$

$\overline{\mathcal{L}}$ tells us that B_i and F_{3j} are canonically conjugate variables, satisfying

$$[B^{ai}(x), \varepsilon_{jj'} F^{a'}_{3j'}(0)]_{x_0=0} \simeq \frac{1}{i} \delta^{aa'} \delta_{ij} \delta^3(x). \qquad (2.16)$$

Now, integrating out $F_{\mu\nu}$ in (2.15), one gets

$$I = \int \mathcal{D}A^{(0)} \mathcal{D}B_\mu \exp\left(i \int d^4x\, \mathcal{L}(B, A^{(0)})\right). \qquad (2.17)$$

$\mathcal{L}(B, A^{(0)})$ is complicated but can be expanded in powers of B:

$$\int d^4x\, \mathcal{L}(B, A^{(0)}) = \frac{1}{2} \int \frac{d^4p}{(2\pi)^4} B^a_\alpha(p) M^{\alpha\beta}_{ab}(p) B^b_\beta(-p) + O(B^3),$$
$$M^{\alpha\beta}_{ab}(p=0) = g^2 f^{acd} f^{bce} (A^{(0)d}_\gamma A^{(0)e\gamma} g^{\alpha\beta} - A^{(0)d\alpha} A^{(0)e\beta}). \qquad (2.18)$$

This mass matrix can be easily shown to be positive semidefinite and moreover, if the eigenvalues of $A^{(0)a}_\alpha$ are all non-zero, it is positive definite. In this way, he obtains

$$\mathcal{L}_{eff}(B) = -\frac{1}{4}(\partial_\alpha B^a_\beta - \partial_\beta B^a_\alpha)^2 - \frac{1}{2}(\partial_3 B^a_\alpha)^2 - \frac{m^2}{2}(B^a_\alpha)^2 + O(B^3), \qquad (2.19)$$

with the propagator

$$i \langle B^a_\alpha B^b_\beta \rangle_p \simeq \delta^{ab} g_{\alpha\beta} \frac{1}{p^2 - m^2} + p_\alpha p_\beta \text{ term} \quad . \qquad (2.20)$$

In terms of the canonically conjugate variable,

$$i \langle F^a_{3\alpha} F^b_{3\beta} \rangle_p \simeq -\delta^{ab} g_{\alpha\beta} \frac{m^2}{p^2 - m^2} + p_\alpha p_\beta \text{ term} . \qquad (2.21)$$

In (2.20) and (2.21), m^2 is given, for SU(N), by

$$\frac{1}{m^2} = \frac{1}{3(N^2-1)} \sum_{\ell=1}^{3(N^2-1)} \frac{1}{\xi_\ell} \quad , \tag{2.22}$$

where ξ_ℓ are the eigenvalues of $M_{ab}^{\alpha\beta}(p=0)$. Thus he arrives at the low energy effective Lagrangian

$$\begin{aligned} \mathcal{L}_{eff} &= -\frac{1}{2}\partial_\mu \phi_\alpha^a \partial^\mu \phi^{a\alpha} + \frac{1}{2} m^2 \phi_\alpha^a \phi^{a\alpha} + \cdots \quad , \\ \phi_\alpha^a &= F_{3\alpha}^a / m \ . \end{aligned} \tag{2.23}$$

In other words, due to the gluon condensation, the field strength $F_{3\alpha}^a$ and the dual vector potential B_α^a become massive. As B_α^a is what couples to the color magnetic current, this means that the system is in the magnetically superconducting phase (or magnetic Higgs phase). Then one can deduce the area decay for the Wilson loop in the following two ways.

(a) Take the abelian Wilson loop W(C) and the dual loop M(C').

$$\begin{aligned} W(C) &= \exp ig \oint_C A_\mu^3 dx^\mu \quad , \\ M(C') &= \exp ig' \oint_{C'} B_\mu^3 dx^\mu \ . \end{aligned} \tag{2.24}$$

By using the equal time commutation relations, one can derive the 't Hooft equation [7]

$$W(C)M(C') = M(C')W(C)\exp igg'n \quad , \tag{2.25}$$

where n is the number of times C' winds around C. As B is massive, M(C') shows the perimeter law and since there is no massless excitation at low energy, W(C) should obey the area law a la 't Hooft.

(b) Alternatively, one can compute how the electric field decreases when one goes away from the plane of a Wilson loop. Taking the loop to be in 0-3 plane, and solving the equation of motion for F_{30}^3, one gets

$$F_{30}^{3} = -\frac{1}{2\pi} m^2 g K_0(mr) \ , \ K_0(x) = \text{modified Bessel function}, \\ r = \text{perpendicular distance from the plane}. \tag{2.26}$$

Thus the electric flux is confined on the plane within the thickness $\sim 1/m$. So, although some assumptions had to be made, Fukuda's analysis nicely shows how the gluon condensation is related to confinement.

§3. Gluon Condensation in Super QCD and a Puzzle

The rest of the talk is based on a recent work done in collaboration with H. Hata [8].

Having seen that everything seems to be O.K. in QCD with $\langle 0| F_{\mu\nu}^2 |0\rangle > 0$, let us now turn to supersymmetric QCD. As we have been told thousands of times, supersymmetry must be broken to be useful and it should occur naturally. At present, this means either it's breaking scale is induced by the Planck scale or the breaking is dynamical, the scale being produced by dimensional transmutation. We will exclusively deal with the latter possibility.

In four dimensions, dynamical supersymmetry breaking has been prohibitively difficult to achieve. As we know, Witten has given an ingenious reason for that based on the concept of the index Δ of a supersymmetric theory [9]. It is defined by

$$\Delta = \mathrm{Tr}(-1)^F = n_B^{E=0} - n_F^{E=0} \quad , \tag{3.1}$$

where $n_B^{E=0}$ ($n_F^{E=0}$) is the number of zero energy bosonic (fermionic) states. Δ is, according to Witten, a quantity independent of the parameters of theory such as the coupling constant and the masses as long as one excludes special values of the parameters. If $\Delta=0$, Witten's analysis says nothing, but if it is non-vanishing, then we know that there is at least one state with exactly zero energy and supersymmetry is not broken. For all the four dimensional theories for which Δ has been calculated, it was non zero. For example, for SU(N) super QCD, $\Delta = N$.

Powerful as it is, Witten's analysis certainly leaves room for further study.

(i) First, to me, the argument for the independnece of Δ on the parameters of the theory seems formal and not water-tight. In particular,

exactly massless case might be suspected.
(ii) By its very nature, dynamical understanding is not gained. Even for theories with $\Delta \neq 0$, I believe it's important to unravel why and how supersymmetry is unbroken, for it may give valuable insight for the case where $\Delta = 0$ or Δ is not reliably calculable.

With this motivation, let us make an attmept to study SU(N) super QCD dynamically. The Lagrangian in the Wess-Zumino gauge, after eliminating the auxiliary field, is of the familiar form

$$\mathcal{L} = -\frac{1}{4} F^a_{\mu\nu} F^{\mu\nu a} + \frac{1}{2} \bar{\lambda} i \not{D} \lambda \quad , \tag{3.2}$$

where λ is a Majorana spinor in the adjoint representation. The theory is asymptotically free with

$$\beta(g) = -3N \frac{g^3}{16\pi^2} + \cdots \quad . \tag{3.3}$$

For this theory, the order parameter for the supersymmetry breaking is the gluon condensation $\langle 0|F^2_{\mu\nu}|0\rangle$, which is easily seen again from the trace anomaly equation

$$\begin{aligned} \mathcal{E}_{vac} = \frac{1}{4}\langle 0|\theta^\mu_\mu|0\rangle &= -\frac{\beta(g)}{2g}\langle 0|(-\frac{1}{4}F^2_{\mu\nu} + \frac{3}{4}\frac{1}{2} i \bar{\lambda}\not{D}\lambda)|0\rangle \\ &= \frac{\beta(g)}{8g}\langle 0|N(F^2_{\mu\nu})|0\rangle . \end{aligned} \tag{3.4}$$

Here, by $N(F^2_{\mu\nu})$ it is indicated that the quartic divergence in the vacuum expectation value of $F^2_{\mu\nu}$ is precisely canceled by that of $\bar{\lambda}\not{D}\lambda$, which is otherwise a null operator and vanishes in the vacuum.

Now let us apply the same analysis as before utilizing the trace anomaly equation in the presence of a constant source for $\mathcal{L}$. We do not have to repeat the calculation. The only difference is the slight change in the β function of the theory, which nevertheless guarantees asymptotic freedom. At this point we realize that we are faced with a puzzle; $N(F^2_{\mu\nu})$ condenses with positive sign so that $\mathcal{E}_{vac}$ becomes negative. Supersymmetry gets broken, and does so in a pathological manner. I.e., a Nambu-Goldstone fermion $F^a_{\mu\nu}\sigma^{\mu\nu}\lambda^a$ appears endowed with negative norm. Everything that looked natural and desirable in ordinary QCD now tells us that supersymmetric QCD is a sick theory!

Before examining this peculiar circumstance more closely, let me make three remarks.

(i) This type of pathology was observed before in Zanon's model [10] by several authors [11]. It is an O(N)-symmetric Wess-zumino type model in four dimensions with the Lagrangian

$$\mathcal{L} = \frac{1}{4}\int d^4\theta \,(N\phi_0^+\phi_0 + \phi_i^+\phi_i) + \frac{m}{4}\left(\int d^2\theta\, N\phi_0^2 + h.c.\right) + \frac{g}{\sqrt{2}}\left(\int d^2\theta\, \phi_0\phi_i^2 + h.c.\right) , \tag{3.5}$$

where ϕ_0 and ϕ_i are, respectively, O(N) singlet and O(N) vector chiral superfields. In the large N limit the effective potential takes the form

$$V_{eff} = \frac{N\mu^4}{16\pi^2}\Big\{-\frac{2\pi^2}{g^2}f^2 - \frac{4\pi^2}{g^2}\frac{m}{\mu}af + \frac{1}{4}\big[(a^2+f)^2\ln(a^2+f) + (a^2-f)^2\ln(a^2-f) - 2a^4\ln a^2 - 3f^2\big]\Big\} , \tag{3.6}$$

where $a = 2gA_0/\mu$, $f = 2gF_0/\mu^2$ and μ is the renormalization point. As is depicted in Fig. 2, the vacuum energy is negative at the stationary point and there indeed exists a negatively-normed Nambu-Goldstone fermion. This pathology however is caused through renormalization and is due to the lack of asymptotic freedom. On the contrary, the pathology we encounterd above is <u>because of</u> the asymptotic freedom.

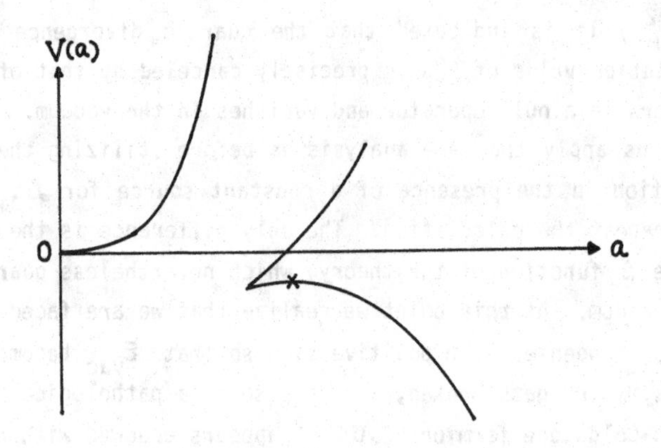

Fig. 2. Effective potential V(a) for Zanon's model.

(ii) The pathology cannot be resolved by Witten's index argument. First let us see how bosonic and fermionic states are paired for negative energy density. If we take $|b\rangle$ to be a normalized bosonic state with energy $E(<0)$, i.e., $\langle b|b\rangle=1$, $H|b\rangle=E|b\rangle$, $H=Q^2$, then there exists a fermionic state $|f\rangle \equiv Q/\sqrt{E}|b\rangle$ which is degenerate with $|b\rangle$,

$$H|f\rangle = \frac{Q}{\sqrt{E}}H|b\rangle = E\frac{Q}{\sqrt{E}}|b\rangle = E|f\rangle \; . \tag{3.7}$$

It's norm, however is negative,

$$\langle f|f\rangle = \langle b|\frac{Q}{\sqrt{E}^*}\frac{Q}{\sqrt{E}}|b\rangle = \frac{E}{|E|} = -1 \; . \tag{3.8}$$

This is consistent with the fact that we must have a Nambu-Goldstone fermion with negative metric. Now we can make the point. The point is that Witten's argument presupposes $E \gtrsim 0$. If we include the possibility of $E<0$, we actually have five cases schematically shown in Fig. 3.

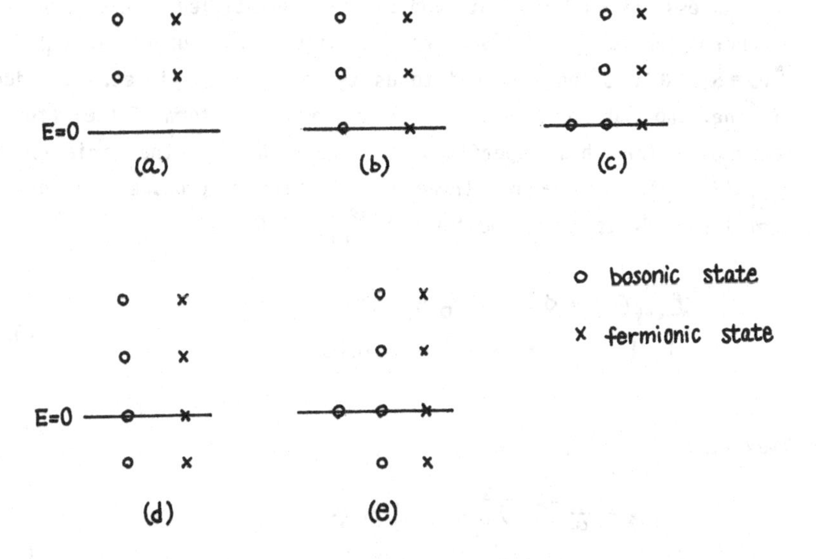

Fig. 3. Five possibilities for the configuration of bosonic and fermionic states in supersymmetric theories.

In particular, look at the possibility (e). Δ, which concerns only the states with E=0, may be counted by perturbation theory as Witten does, but the states with E<0 are of non-perturbative origin and cannot be assessed by perturbative means. In (e) clearly $\Delta \neq 0$ but supersymmetry is broken. Thus our pathology is <u>not inconsistent</u> with non-vanishing index.

(iii) The third remark concerns the recent work of Veneziano and Yankielowicz [12], who tried to construct the effective Lagrangian for super QCD utilizing the anomaly structure of the theory. It sounds very similar to what I have been discussing, but it is actually entirely different. So it may be helpful to clarify this point. As is well known, in N=1 super Yang-Mills theory, currents associated with the superconformal, dilatation, and chiral transformations have anomalies, which are related.

$$\begin{aligned} \partial_\mu (x_\nu \gamma^\nu S^\mu) = \gamma_\mu S^\mu &\propto \frac{\beta(g)}{2g} F^a_{\mu\nu} \sigma^{\mu\nu} \lambda^a \\ \partial_\mu d^\mu = \partial_\mu (x_\nu \theta^{\mu\nu}) = \theta^\mu_\mu &\propto \frac{\beta(g)}{2g} F^a_{\mu\nu} F^{\mu\nu a} \\ \partial_\mu j^{\mu 5} &\propto \frac{\beta(g)}{2g} F^a_{\mu\nu} \tilde{F}^{\mu\nu a} \end{aligned} \tag{3.9}$$

The expressions on the right hand side are identified as components of a renormalization group invariant composite chiral superfield ($\beta(g)/g$) $W^\alpha W_\alpha \equiv S(x,\theta)$, often referred to as the anomaly multiplet. The idea of Veneziano and Yankielowicz is to determine the form of the effective Lagrangian for this superfield by requiring that the variation of $\mathcal{L}_{eff}(S)$ under the above three transformations produce the correct anomalies. By assuming the form of $\mathcal{L}_{eff}(S)$ to be

$$\begin{aligned} \mathcal{L}_{eff}(S) &= d(S,S^*)_D + f(S)_F + h.c. \\ d,\ f &= \text{arbitrary functions} \ , \end{aligned} \tag{3.10}$$

they find

$$\begin{aligned} \mathcal{L}_{eff} &= \frac{g}{\alpha}(S^\dagger S)^{1/3}_D + \frac{1}{3}\left(S \ln \frac{S}{\mu^3} - S + h.c.\right)_F \\ V_{eff} &= -\frac{1}{3}\left(S \ln \frac{S}{\mu^3} - S + h.c.\right)_F \ . \end{aligned} \tag{3.11}$$

Although anomalies are correctly reproduced, the ansatz is clearly not the most general one. For instance, V_{eff} contains the F component of S(i.e. $F_{\mu\nu}^2$) only linearly. The term such as $F_{\mu\nu}^2 \ln F_{\mu\nu}^2/\mu^4$, which we expect to find, can only be obtained by an expression like $[S \ln \bar{D}DS]_F$, involving covariant derivatives as well. In fact D'Adda, Davis, Di Vecchia and Salomonson [13] recently constructed $\mathcal{L}_{eff}$ for supersymmetric CP^{n-1} model explicitly and found terms of $\bar{D}DS$ type, which cannot be fixed by the requirement of anomaly reproduction. While the approach of Veneziano and Yankielowicz uses anomaly equations kinematically as Ward-Takahashi identities, what I described before was to use the trace anomaly equation more dynamically by converting it into a non-linear differential equation for V_{eff}. These approaches are quite different.

§4. The Puzzle in Supersymmetric Non-linear σ Model in Two Dimensions and How the Puzzle is Resolved

Let us now go back to the puzzle. By itself the puzzle seems quite difficult to solve. But as is often the case, we can get a valuable insight by studying a simpler solvable model which exhibits the same phenomenon. Supersymmetric O(N) non-linear σ model in two dimensions is just such a model. As is well-known the ordinary (i.e. non-super) O(N) non-linear σ model has many important features in common with four dimensional Yang-Mills theory. The model is asymtotically free, formally conformally invariant with anomaly at the quantum level and, in addition, as has recently be pointed out by Hata [14], the way the symmetry is realized in the true vacuum seems to be very similar to the case of Yang-Mills thoery. In the σ model, O(N) symmetry, which is spontaneously broken down to O(N-1) in perturbation theory, is restored by disordering of the fields and realized linearly in the large N true vacuum. In Yang-Mills theory, a special residual gauge symmetry within the Lorentz gauge is spontaneously broken in perturbative vacuum (gluons are the associated Nambu-Goldstone bosons) and as was shown by Hata, the condition for the restoration of this symmetry coincides with the condition for color confinement [15].

In any case what is important at the moment is that, as we shall see, a similar puzzle exists in the supersymmetric version of this model and we can completely analyze this in the large N limit. As has already been discussed by many people [16], the model does not break supersymmetry. What we will focus on is <u>how</u> this happens and what it teaches us about the puzzle in super QCD.

The model is defined by the action

$$S = \int d^2x \int d^2\theta \frac{1}{2\lambda_0}\left[\bar{D}_\alpha \Phi_0^i D_\alpha \Phi_0^i - 2\phi_0(\Phi_0^i \Phi_0^i - 1)\right] \tag{4.1}$$

where ϕ_0 and Φ_0^i are respectively O(N) singlet and O(N) vector superfields with components

$$\begin{aligned} \Phi^i &= A^i + \bar{\theta}\psi^i + \tfrac{1}{2}\bar{\theta}\theta F^i \ , \\ \phi &= a + \bar{\theta}\chi + \tfrac{1}{2}\bar{\theta}\theta f \ . \end{aligned} \tag{4.2}$$

In terms of components the Lagrangian is of the form

$$\begin{aligned} \mathcal{L} = \frac{1}{2\lambda_0}\{ & (\partial_\mu A_0^i)^2 + i\bar{\Psi}_0^i \not{\partial} \Psi_0^i + F_0^i F_0^i \\ & - 2(a_0 F_0^i A_0^i - a_0 \bar{\Psi}_0^i \Psi_0^i - 2\bar{\chi}_0 A_0^i \Psi_0^i) - f_0 (A_0^{i\,2} - 1)\} . \end{aligned} \tag{4.3}$$

By the use of equations of motion, we may also write it in the form

$$\left\{ \begin{aligned} & \mathcal{L} = \frac{1}{2\lambda_0}[(\partial A_0^i)^2 + i\bar{\Psi}_0^i \not{\partial} \Psi_0^i + (\tfrac{1}{2}\bar{\Psi}_0^i \Psi_0^i)^2] \ , \\ & A_0^i \Psi_0^i = 0 \ , \\ & A_0^i A_0^i = 1 \ . \end{aligned} \right. \tag{4.4}$$

It is a sum of the non-linear σ model and the Gross-Neveu model [17] with common coupling g and a constraint. The renormalization of the model is very simple

$$\begin{aligned} & A^i = Z^{1/2} A_0^i \ , \quad \psi^i = Z^{1/2}\psi_0^i \ , \quad \lambda_0 = \mu^\varepsilon Z \lambda \ , \\ & \text{Others not renormalized} \ , \\ & \frac{1}{Z} = 1 + \frac{N\lambda}{4\pi}\left(\frac{2}{\varepsilon} - \gamma_E + \ln 4\pi\right) \ , \end{aligned} \tag{4.5}$$

where the dimensional reduction regularization [18] with dim= ν=2- ε is employed. The β function in the large N limit is given by

$$\beta(\lambda) = -\frac{N\lambda^2}{2\pi} < 0. \tag{4.6}$$

First let us recall the main features of the ordinary non-linear σ model, which is obtained by retaining only A^i and f in the above Lagrangian. The effective potential for the variables f and A^i is shown in Fig. 4. (As can be easily shown, condensation of f is equivalent to the condensation of the Lagrangian operator.) The stationary

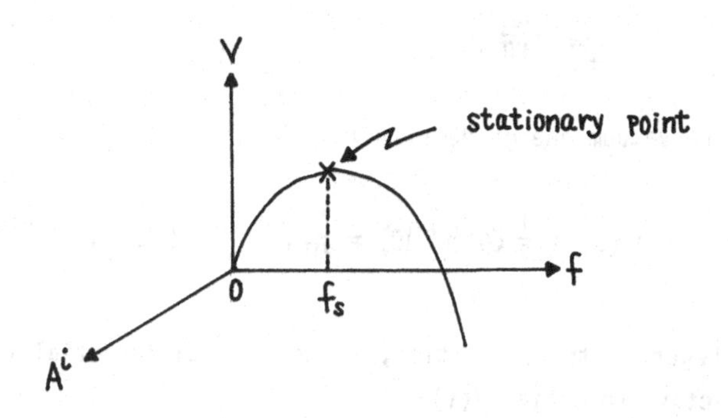

Fig. 4. Effective potential $V(f, A^i)$ for O(N) non-linear σ model.

point is at $(f, A^i)=(f_s, 0)$ and $V(f_s, 0)>0$. The O(N) vector fields acquire a mass gap, $m^2_{Ai}=f_s$. Although the vacuum is at the saddle point of V_{eff}, it is not a problem: One can easily show that there is no particle pole in the f channel, and the unboundedness of V_{eff} does not mean that the energy is unbounded. (As was discussed by Coleman [19], for a composite operator such as f, V_{eff} is interpretable as an energy density only at its stationary point.)

Now let us look at the trace anomaly equation for the supersymmetric case. Just as in QCD case, we introduce the source J for the Lagrangian operator, which we can show will not be renormalized,

$$\begin{aligned}\mathcal{L}_J &= \left(\frac{1}{2\lambda_0}+J\right)\left[(\partial A_i)^2 + i\bar{\Psi}_i \not\partial \Psi_i + \left(\tfrac{1}{2}\bar{\Psi}_i\Psi_i\right)^2\right] \\ &\equiv \left(\frac{1}{2\lambda_0}+J\right)\hat{f} \quad .\end{aligned} \tag{4.7}$$

The energy momentum tensor is of the form

$$\theta_{\mu\nu} = \left(\frac{1}{\lambda_0} + 2J\right)\left\{\partial_\mu A_i \partial_\nu A_i + \frac{i}{4}\left(\bar{\Psi}_i \gamma_\mu \partial_\nu \Psi_i + (\mu \leftrightarrow \nu)\right) - \eta_{\mu\nu}\frac{1}{2}\hat{f}\right\}. \tag{4.8}$$

Taking the trace in 2-ε dimension, we get

$$\theta_\mu^{\ \mu} = \varepsilon\left(\frac{1}{2\lambda_0} + J\right)\left\{(\partial A_i)^2 + i\bar{\Psi}_i \not{\partial} \Psi_i + \left(\frac{1}{2}\bar{\Psi}_i \Psi_i\right)^2\right\} \underset{\varepsilon \to 0}{=} \frac{N}{4\pi}\hat{f} \ . \tag{4.9}$$

So the vacuum energy density is

$$\varepsilon(J) = \frac{1}{2}\langle 0|\theta_\mu^{\ \mu}|0\rangle = \frac{N}{8\pi}f \quad , \quad f = \langle 0|\hat{f}|0\rangle . \tag{4.10}$$

By Legendre transformation, we get the differential equation for the effective potential V(f)

$$V(f) = \varepsilon(J) - J\frac{d\varepsilon}{dJ} = \frac{N}{4\pi}f + f\frac{dV}{df} \ . \qquad \left(\frac{d\varepsilon}{dJ} = f \ , \ \frac{dV}{df} = -J\right) \tag{4.11}$$

This is easily solved to give

$$V(f) = -\frac{N}{8\pi}f\left(\ln\left(\frac{\pm f}{\mu^2}\right) + \text{const.}\right) . \tag{4.12}$$

Note that this is of the same form as in super QCD and we get the same puzzle. This time let us keep the two cases, $f \geq 0$ and $f \leq 0$. The potential is shown in Fig. 5. In either case, $\varepsilon_{vac} \neq 0$ and supersymmetry must be broken, which we know is false. What went wrong?

The answer is provided when we consider another important order parameter of the theory

$$a \equiv \langle 0|\frac{1}{2}\bar{\Psi}_i \Psi_i|0\rangle \quad , \tag{4.13}$$

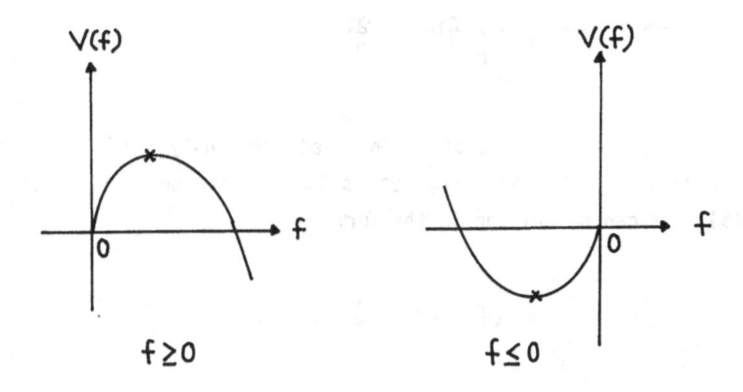

Fig. 5. Two possibilities of the effective potential V(f), obtained via trace anomaly, for supersymmetric O(N) non-linear σ model.

together with f. If we introduce the source for $\frac{1}{2}\bar{\Psi}_i\Psi_i$ and rederive the trace anomaly equation, following the steps similar to the one before, we find

$$V(f,a) = \frac{N}{8\pi}f + f\frac{\partial V}{\partial f} + \frac{1}{2}a\frac{\partial V}{\partial a} \ . \tag{4.14}$$

For convenience, let us rewrite this in terms of redefined dimensionless variables

$$\frac{N}{8\pi}\frac{f}{M^2} \longrightarrow f \ , \qquad M = \mu \exp\left(-\int_{\lambda_0}^{\lambda}\frac{d\lambda'}{\beta(\lambda')}\right)$$
$$\frac{a}{M} \longrightarrow a \ , \qquad = \text{renormalization group}$$
$$\frac{V}{M^2} \longrightarrow v \ , \qquad \text{invariant mass.}$$

Then we have

$$v = f + f\frac{\partial v}{\partial f} + \frac{1}{2}a\frac{\partial v}{\partial a} \ . \tag{4.15}$$

Normally, we say that this is an unnecessary step, since at the stationary point, $\partial v/\partial a=0$, the additional term vanishes, and

$$\frac{dv}{df} = \frac{\partial v}{\partial f} + \frac{\partial a}{\partial f}\frac{\partial v}{\partial a} = \frac{\partial v}{\partial f} \quad , \tag{4.16}$$

i.e., we get back the equation obtained previously. But here it is not so trivial. To see this, let us look at the general solution of (4.15). It can be written in the form

$$\begin{cases} v = -f\ln(f/e) + f\,\Phi(x) \\ x \equiv a^2/f \\ \Phi(x) = \text{arbitrary function.} \end{cases} \tag{4.17}$$

The point is that the elimination of a by $\partial v/\partial a=0$ is non-trivial. In fact $0=\partial v/\partial a=2a\,\Phi'(x)$ and we have at least two solutions,

$$\begin{cases} a = 0 & \text{(A)} \\ \Phi'(x) = 0 & \text{(B)} \end{cases} \tag{4.18}$$

In case (A), we go back to the previous equation $v=f+dv/df$ and we have $f_s=\pm\exp\Phi(0)\neq 0$. Supersymmetry is broken.

But we also have the possibility (B). There are two cases within case (B). (Bi) Suppose $\Phi'(x)=0$ has a solution at finite x, call it x_s. Then, $v=-f\ln(f/e)+f\Phi(x_s)$ and we find $f_s=\pm\exp\Phi(x_s)\neq 0$. Again supersymmetry is broken. (Bii) The second case appears if $v(a,f)\to 0$ as $f\to 0$. In this case, it is clear that $\Phi(x)$ must behave like x^{ν} for large x (i.e. for small f) with $\nu<1$. Then $\Phi'(x)\sim x^{\nu-1}\to 0$ and $x=\infty$, or $f=0$, is a root of $\Phi'(x)=0$ for any finite value of a. Thus just by solving $\partial v/\partial a=0$ we automatically get $f=0$, $v=0$. Supersymmetry is unbroken. To be precise, there is one more condition. For this to be a stationary point, it must of course satisfy $\partial v/\partial f=0$ as well. It can easily be shown that this requires $\Phi(x)\sim -\ln x$ for large x. Explicit large N calculation shows [16] that indeed the case (Bii) is realized; $\Phi(x)$ is of the form

$$\Phi(x) = x \ln x - (1+x)\ln(1+x) \ , \tag{4.19}$$

and $\Phi'(x)=\ln(1+1/x)=0$ has a unique solution at $x=\infty$.

We have learned a lesson: The lesson is that it is sometimes dangerous to assume that introducing a source and then eliminating it by its stationarity condition is the same as not introducing the source at all. The puzzle in super non-linear σ model is resolved precisely by this subtlety.

Now let us go back to super QCD and look at the trace anomaly equation with the source term $-m/2\,\bar{\lambda}\lambda$. The energy density now becomes

$$\begin{aligned} &\mathcal{E}(J,m) = -\frac{\beta(g_J)}{2g_J}(1+J)F - \frac{1}{4}(1+\gamma_m(g_J))mA \ , \\ &F \equiv \langle 0| -\tfrac{1}{4}F_{\mu\nu}^2 |0\rangle \quad , \quad A \equiv \langle 0| \tfrac{1}{2}\bar{\lambda}\lambda |0\rangle \ , \\ &\gamma_m(g) = \text{anomalous dimension for } \bar{\lambda}\lambda \ . \end{aligned} \tag{4.20}$$

For qualitative understanding, we may take g to be small. Then by Legndre transformation we get $V=-(b_0/2)g^2F+F\,\partial V/\partial F+(3/4)A\,\partial V/\partial A$. Rescaling by $v\equiv V/M^4$, $f\equiv -(b_0/2)g^2F/M^4$, $a\equiv (A/M^3)^{2/3}$ brings this to the form

$$v = f + f\frac{\partial v}{\partial f} + \frac{1}{2}a\frac{\partial v}{\partial a} \ , \tag{4.21}$$

which is identical to the equation for super non-linear σ model. Thus it is quite possible that by the same mechanism as before, supersymmetric stationary point $f=0$, $a\neq 0$, $v=0$ can exist for super QCD.

§5. Remainder of the Puzzle and the Picture of Confinement in QCD and Super QCD

We have seen that by considering $\langle 0|\bar{\lambda}\lambda|0\rangle$ together with $\langle 0|F_{\mu\nu}^2|0\rangle$, it is possible to avoid overlooking the supersymmetric solution $\langle 0|F_{\mu\nu}^2|0\rangle=0$. But this alone does not solve the puzzle entirely. As concerns the shape of the effective potential $V(A, F)$, there are two possibilities dipicted in Fig. 6. Let us discuss these cases separately. Case (A); The puzzle is not yet resolved. The pathological stationary

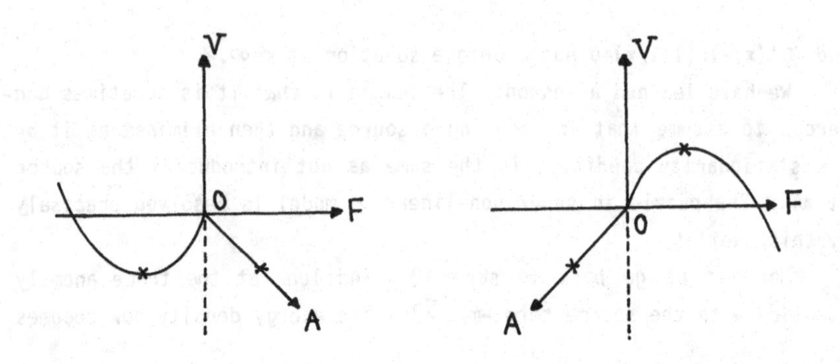

Fig. 6. Two possibilities of the effective potential V(A, F) for supersymmetric QCD.

point is still lower in energy compared with the supersymmetric one and we must show that the former is unstable. Moreover such instability should occur exclusively for the supersymmetric case, for otherwise, the nice picture of ordinary QCD, with the vacuum energy lower than the perturbative one, will be lost. At this moment, we can only suggest a possible scenario. In supersymmetric theories, at the tree level, there is a famous mass formula. For a chiral multiplet, for expample, $m_A^2+m_B^2=2m_\psi^2$, where m_A and m_B are the masses of the scalar and the pseudoscalar. For the effective Lagrangian, this formula is not expected to hold exactly. But suppose that the deviation is not very large. Then, applying this to our composite chiral multiplet $W^\alpha W_\alpha$, in case (A) the right hand side vanishes because ψ is the Nambu-Goldstone fermion. Then unless there is a hidden symmetry which makes both m_A and m_B to be small, it appears that either the scalar or the pseudoscalar becomes tachyonic, giving rise to an instability.

Case(B): This is precisely of the non-linear σ model type. Here the puzzle is completely resolved; the ground state is supersymmetric and the pathology disappears. However, in this case a question immediately arises whether this is consistent with confinement, including the case of non-supersymmetric QCD. For supersymmetric case, $\langle 0|F_{\mu\nu}^2|0\rangle=0$, and for ordinary QCD, $\langle 0|F_{\mu\nu}^2|0\rangle<0$, i.e., the condensation is electric rather than magnetic. In other words, we ask the question " is magnetic condensation necessary for confinement?"

Since the situation will be quite similar in supersymmetric case, let us take the ordinary Yang-Mills theory and consider the Wilson loop in the vacuum. Removing the linear divergence along the perimeter, which is due to the self energy of the infinitely massive test quark, $W(C)$ obeys the renormalization group equation:

$$\left(\mu\frac{\partial}{\partial\mu}+\beta(g)\frac{\partial}{\partial g}\right)\langle W(C)\rangle = 0 \; . \tag{5.1}$$

Taking the loop to be a circle with area S, one can rewrite this equation in the form [3]

$$\begin{aligned} &\left(\frac{\partial}{\partial S}+K(S)\right)\langle W(C)\rangle = 0 \\ &K(S) = -\frac{\beta(g)}{Sg}\int d^4x\, \Delta F(x:C) \quad , \end{aligned} \tag{5.2}$$

wher $\Delta F(x:C)$ is the difference of the condensation of the Lagrangina operator $-\frac{1}{4}F_{\mu\nu}^2$ in the presence and in the absence of the Wilson loop. I.e.,

$$\Delta F(x:C) = \frac{\langle \hat{F}(x)\, W(C)\rangle}{\langle W(C)\rangle} - \langle \hat{F}(x)\rangle \; , \quad \hat{F}(x) = -\frac{1}{4}F_{\mu\nu}^2 \; . \tag{5.3}$$

The point I wish to emphasize is that it is this difference, not the $\langle\hat{F}\rangle$ itself, which is related to confinement. In fact if this difference is uniformly non-vanishing on the plane of a large loop, i.e., if $\int d^4x \Delta F \propto S$ and if it is positive, then, because $\beta(g)<0$, the above differential equation tells us that $\langle W(c)\rangle$ exhibits the area law decay $W(C) \sim \exp(-\alpha S)$. In other words, what is important for confinement is that the response of the gluon condensation to the introduction of the loop is relatively electric and has the support on the plane of the loop.

To convince ourselves of this feature more clearly, we can again use the non-linear σ model as our laboratory. In this two dimensional model the analogue of the Wilson loop is the O(N) invariant two-point correlation function $\langle A_i(x)A_i(0)\rangle$, and the exponential decay for large

$|x|$ corresponds to the area decay of $\langle W(C)\rangle$. As before, the renormalization group equation for $\langle A_i(x)A_i(0)\rangle$ can readily be converted into a differential equation

$$
\begin{aligned}
&(\partial/\partial|x| + M(x))\langle A_i(x)A_i(0)\rangle = 0 \\
&M(x) = -\frac{\beta(\lambda)}{|x|\lambda^2}\int d^2y\, \Delta f(y:x) \qquad (5.4) \\
&\Delta f(y:x) = \frac{\langle \hat{f}(y)A_i(x)A_i(0)\rangle}{\langle A_i(x)A_i(0)\rangle} - \langle \hat{f}(y)\rangle, \quad \hat{f}(y) \equiv (\partial_\mu A_i)^2(y).
\end{aligned}
$$

Remember that $\langle\hat{f}(y)\rangle$ we know is already positive, i.e. "electric" in this model. Since the model is solvable, we can actually compute Δf explicitly for large $|x|$. The result is summarized in Fig. 7. In the

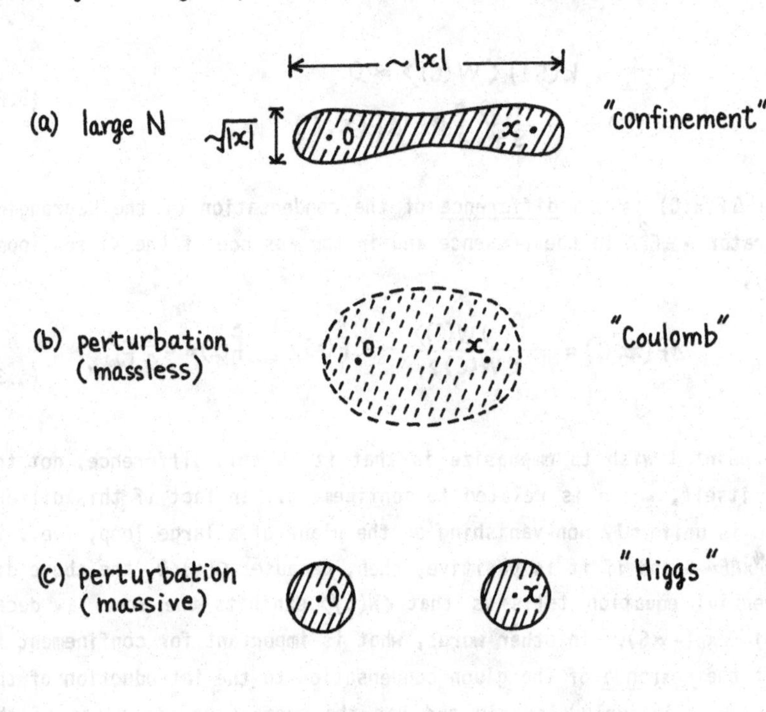

Fig. 7. Support of the function $\Delta f(y:x)$ (explained in the text) calculated in (a) the large N limit, (b) perturbation theory with massless $A^i(x)$, and (c) perturbation theory with massive $A^i(x)$.

large N vacuum, which is considered to be in the correct phase of the theory, indeed the response is relatively electric and is proportional to $|x|$. On the other hand perturbative calculation with massless (Goldstone) and massive A_i (where the mass term is introduced by hand just for the purpose of illustration) show the behavior corresponding to Coulomb and Higgs phase.

We see very clearly that the sign of $\langle\hat{f}\rangle$ itself is not important. Even the magnitude of $\langle\hat{f}\rangle$ is irrelevant: In supersymmetric version of the model, for which $\langle\hat{f}\rangle=0$, we find exactly the same response for Δf, and $\langle A_i(x)A_i(0)\rangle$ exhibits the exponential decay.

Now this observation brings forth a suggestion that also in QCD and super QCD, case (B) might be realized, contrary to the prevailing picture of magnetic gluon condensation. Of course I do not mean to say that such a scenario is more likely. What I do wish to point out is that it is a logical possibility indicated by our study and should be given consideration.

As you may have already been thinking about, there are apparent problems for such a scenario. Let me list them and briefly discuss how serious they actually are.

(i) For QCD, the scenario seems to be in contradiction with the measurement $\langle\frac{\alpha_s}{\pi}F_{\mu\nu}^2\rangle=0.012\ \mathrm{GeV}^4>0$. This is a problem. Assuming the measurement to be correct, the only way out seems to be to argue that the theoretical discussions were made without massive quarks whereas the measurement is made in the region where heavy quarks are important. I simply do not have any better argument for this problem.

(ii) What about the existence of 0^+ glueball? In non-linear σ model, one could check that there was no particle in the relevant channel. But in QCD, does the case (B) type potential not imply tachyonic glueball in $F_{\mu\nu}^2$ channel? It turns out that it is not neccesarily a problem. Firstly, since the effective potential is for p=0, negarive curvature does not directly imply a tachyonic pole. Secondly, because $F_{\mu\nu}^2$ is a hard composite operator, Källen-Lehman representation, which can normally be used to relate the curvature at p=0 to the one at the pole, is not valid for $F_{\mu\nu}^2$. So the existence of 0^+ glueball is not in direct contradiction with the negative curvature in V_{eff}.

(iii) Case (B) scenario indicates that a hadronic bag is filled with electric field. Does it not contradict a fairly successful picture of freely moving quarks and gluons inside the bag? This also is not a serious problem. What should be almost free in the bag is the high frequency modes of the quark and gluon fields. For them, the asymptotic freedom, which <u>is</u> intact, guarantees the absence of strong coupling of these field components to the attendant color electric flux.

(iv) One might also ask the following question: If there is color electric field in the vacuum, as well as in the bag, isn't the vacuum unstable against pair creation of $q\bar{q}$? For this question it is sufficient to recognize that unless the color electric field itself condenses, pair creation does not occur. Condensation of color singlet $(E^a)^2$, which is present even for $\langle F_{\mu\nu}^2 \rangle > 0$, is entirely harmless.

Thus, apart from the first phenomenological problem, which is serious, the other apparent problems can be reconciled or are not actually present. In any case, it is at least very amusing that a picture which is quite unconventional and almost opposite to the usual one nevertheless gives similar physical consequences.

§6. Summary

Now let me briefly summarize what I have discussed. Our prevailing picture of QCD vacuum is that it is a magnetic superconductor having the energy density lower than the perturbative value. This is supported, for example, by the effective potential for $F_{\mu\nu}^2$ constructed from the trace anomaly equation in the presence of the source. When this technique is applied to super QCD, however, the very logic which was quite natural and desirable for QCD lead to a pathology in super QCD: $\mathcal{E}_{vac} < 0$ and unphysical Nambu-Goldstone fermion appears. Similar puzzle occured in super non-linear σ model in two dimensions and there we understood what exactly went on. Another order parameter $\langle \bar{\Psi}_i \Psi_i \rangle$ was of great importance and we learned a lesson that in some cases elimination of a variable in the effective potential is quite tricky.

Seeing that super QCD has quite an analogous structure, we were then faced with two possibilities. If one takes the choice which seems to be smoothly connected with the conventional picture of magnetic gluon condensation in QCD, the pathology still persists. A mass formula might solve the puzzle by indicating an instability of such a vacuum, but it remains to be demonstrated.

On the other hand if one takes the alternative possibility, which is of super non-linear σ model type, the puzzle is resolved with the ground state being supersymmetric. We saw that it is consistent with confinement. In fact we found that it is the relatively electric nature of the <u>response</u> of the gluon condensation to the introduction of the test charge, not the nature of the gluon condensation itself, that distinguishes the phases of the system.

This then lead to a logical possibility of an unconventional picture of QCD vacuum, i.e., with "electric" gluon condensation. Except for the conflict with the QCD sum rule result, this scenario, somewhat surprisingly, indicates physics similar to the conventional picture with "magnetic" gluon condensation.

Clearly challange still remains to give unambiguous answers to the problems posed by our comparative study of QCD and supersymmetric QCD.

ACKNOWLEDGEMENT

I would like to thank K. Milton and M. Samuel for their warm hospitality and the well-organized workshop. I am also grateful to H. Hata for useful discussions.

REFERENCES

[1] M.A. Shifman, A.I. Vainstein, and V.I. Zakharov, Pis'ma Zh. Eksp. Teor. Fiz. 27 (1978) 60 JETP Lett. 27 (1978) 55 ; Nucl. Phys. B147 (1979) 385, 448, 519.

[2] G.K. Savvidy, Phys. Lett. 71B (1977) 133. S.G. Matinyan and G.K. Savvidy, Nucl. Phys. B134 (1978) 539. N.K. Nielsen and P. Olesen, Nucl. Phys. B144 (1978) 376. H. Pagels and E. Tomboulis, Nucl. Phys. B143 (1978) 485. R. Weder, Phys. Lett. 85B (1979) 249. R. Fukuda, Phys. Rev. D21 (1980) 485. R. Fukuda and Y. Kazama, Phys. Rev. Lett. 45 (1980) 1142. P. Minkowski, Nucl. Phys. B177 (1981) 203. K. Olaussen and F. Ravndal, Phys. Lett. 100B (1981) 497. K.A. Milton, Phys. Lett. 104B (1981) 49.

[3] R. Fukuda and Y. Kazama, Phys. Rev. Lett. 45 (1980) 1142. See also R. Fukuda, Phys. Rev. D21 (1980) 485.

[4] J.C. Collins, A. Duncan, and S.D. Joglekar, Phys. Rev. D16 (1977) 438.

[5] R. Fukuda, Prog. Theor. Phys. 67 (1982) 648, 655; 68 (1982) 602.

[6] M.B. Halpern, Phys. Rev. D19 (1979) 517.

[7] G 't Hooft, Nucl. Phys. B138 (1978) 1.

[8] H. Hata and Y. Kazama, in preparation.

[9] E. Witten, Nucl. Phys. B202 (1982) 253.

[10] D. Zanon, Phys. Lett. 104B (1981) 127.

[11] A. Higuchi and Y. Kazama, Nucl. Phys. B206 (1982) 152. P. Salomonson, Nucl. Phys. B207 (1982) 350. D. Amati and K.C. Chou, Phys. Lett. 114B (1982) 129.

[12] G. Veneziano and S. Yankielowicz, Phys. Lett. 113B (1982) 321.
[13] A. D'Adda, A.C. Davis, P. DiVecchia and P. Salomonson, CERN preprint (1982).
[14] H. Hata, Prog. Theor. Phys. 67 (1982) 1607.
[15] T. Kugo and I. Ojima, Prog. Theor. Phys. Suppl. 66 (1979) 1.
[16] E. Witten, Phys. Rev. D16 (1977) 2991. P. DiVecchia and S. Ferrara, Nucl. Phys. B130 (1977) 93. O. Alvarez, Phys. Rev. D17 (1978) 1123. E. Witten, Nucl. Phys. B149 (1979) 3235. A.C. Davis, P. Salomonson and J.W. van Holten, Phys. Lett. 113B (1982) 472; Nucl. Phys. B208 (1982) 484. K. Higashijima and T. Uematsu, Phys. Lett. 123B (1983) 209.
[17] D. Gross and A. Neveu, Phys. Rev. D10 (1974) 3235.
[18] W. Siegel, Phys. Lett. 84B (1979) 193; 94B (1980) 37.
[19] S. Coleman, Proceedings of the International School of Subnuclear Physics, Erice, Italy, 1979. Ed. A. Zichichi (Plenum Press, New York) p11.

STABILITY OF A CHIRAL BREAKING VACCUUM

Richard W. Haymaker and Juan Perez-Mercader

1. Introduction

We would like to discuss an effective potential approach to dynamical symmetry breaking.[1,2,3] In this approach, stationary points of the effective potential correspond to solutions of a Schwinger-Dyson equation. The second derivatives of the effective potential give a stability condition. We will exhibit a case in which a symmetry breaking solution of the Schwinger-Dyson equation corresponds to a saddle point of the effective potential and hence is based on a presumed vacuum state that is unstable.[1]

The effective potential is a function that is useful in studying the stability of the ground state of a system. One introduces sources that break symmetries explicitly, calculates the ground state energy in the presence of these external sources, and then Legendre transforms to get the effective potential. If one couples an external source J to the system through an interaction $-J\phi$ then the effective potential $V(\phi)$ is the work done by the external source in displacing the field to the value ϕ. The work done by an external source must be positive in order for the system to be stable. The effective potential formalism generalizes immediately to fields coupled bilinearly to sources which is applicable to the problem of dynamical symmetry breaking by condensates with the appropriate quantum numbers.

We are interested in a tractable first approximation to dynamical symmetry breaking with applications to the problem of chiral breaking in QCD and the patterns of symmetry breaking in Grand Unified Theories. We show that there exists an effective potential $V(\phi_i)$ where the ϕ_i correspond to the spectrum of a Bethe-Salpeter kernel. We introduce a complete set of sources coupled to bilinear combinations of the fields and we exploit this generality showing that it simplifies the Legendre transform. For a non-trivial approximation, the effective potential is calculable in closed form as we will show and further it exhibits an unstable stationary point.

In the following sections we first introduce appropriate sources, (Sec. 2). Then we look at the free field case, defining the ground state energy as a function of these sources and Legendre transforming to get the effective potential, (Sec. 3). We then introduce interactions through a truncated Schwinger-Dyson equation and repeat the steps for the free field, (Sec. 4). Finally we give an example which is worked out in closed form illustrating a symmetry breaking solution that is a saddle point of the effective potential, (Sec. 5), and finish by presenting some conclusions (Sec. 6).

2. Sources

Consider a field theory involving fermions and add to the action an external source term of the form:

$$\int d^4x \int d^4y \bar{\psi}(x)\tilde{J}(x,y)\psi(y). \tag{2.1}$$

$\tilde{J}$ is a translation invariant matrix valued function of the single four-vector $(x - y)^\mu$. Take its Fourier transform and expand it in a complete set of matrix valued functions of p^μ,

$$J(p) = \sum_n J_n F_n(p). \tag{2.2}$$

This leads to zero momentum insertions on fermion lines of the form $S(p)\, F_n(p)\, S(p)$. At this point the particular choice of basis functions is arbitrary. The choice of coupling sources of bilinears in the fields was made in order to be sensitive to instabilities due to tachyonic bound states.

With the sources on, one can define the generating function in the usual way:

$$Z = \exp\left(iW \int d^4x\right) = \int [d\phi]e^{iA}. \tag{2.3}$$

Here a four dimensional volume factor was taken out because the sources are translation invariant. W is an ordinary function (rather than a functional) of the set of variables J_n .

3. Free Field

For a free field, the path integral, Eqn. (2.3), can be easily evaluated to give

$$W(J)_{free} = -i \int (dk) \mathrm{Tr} \ln (S^{-1}) \ . \tag{3.1}$$

where $(dk) = \frac{d^4 k}{(2\pi)^4}$ and

$$S^{-1}(p) = \gamma\cdot k - \sum_n J_n F_n(k) \quad . \tag{3.2}$$

This is the generating function for graphs with one closed fermion loop, for example:

$$\frac{\partial^n W}{\dots \partial J_a} = \tag{3.3}$$

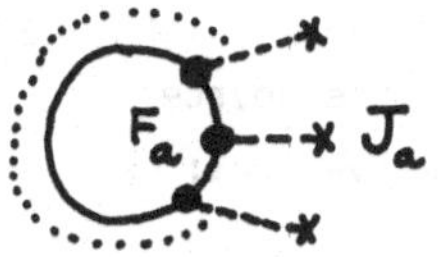

To get the effective potential, first define the variable conjugate to J_n:

$$\phi_n = \frac{\partial W}{\partial J_n} = \frac{\langle 0| \int \int d^4x d^4y \bar{\psi}(x) \tilde{F}_n(x-y) \psi(y) |0\rangle}{(\int d^4 z) \langle 0|0\rangle} \quad . \tag{3.4}$$

The Legendre transform is defined as

$$V(\phi_n) = \sum_m J_m \phi_m - W(J_n) \tag{3.5}$$

If one expands Eqn. (3.5) about the point $\phi_n = 0$, $J_n = 0$, one quickly sees that the calculation pyramids order by order. However, if one has a complete set of functions satisfying the following orthogonality and completeness

$$i \int (dk) \mathrm{Tr}[F_a(k) S_o(k) F_b(k) S_o(k)] = \mu^2 \delta_{ab} \tag{3.6a}$$

$$\sum_a [F_a(k)]_{\alpha\beta} [S_o(p) F_a(p) S_o(p)]_{\gamma\delta}$$
$$= -i(2\pi)^4 \mu^2 \delta_{\beta\gamma} \delta_{\alpha\delta} \delta^{(4)}(k-p) \tag{3.6b}$$

then the calculation collapses to a very simple result:

$$V(\phi)_{free} = + \frac{i}{\mu^2} \sum_a \phi_a \int (dk) \mathrm{Tr}\{\not{k} S_0(k) F_a(k) S_0(k)\} \; +$$

$$-i \int (dk) \mathrm{Tr}\{\ln[\frac{1}{\mu^2} \sum_a \phi_a F_a(k) S_0(k)]\} \quad . \tag{3.7}$$

We used the relation

$$\frac{1}{\mu^2} \sum_b \phi_b F_b = S_0^{-1} S S_0^{-1} \quad , \tag{3.8}$$

where $S_0 = S_{J=0}$, and introduced as arbitrary mass scale μ^2.

The Legendre transform property

$$\frac{\partial V}{\partial \phi_n} = J_n \tag{3.9}$$

assures that V is stationary when the sources are off. The stability is determined by the matrix of derivatives:

$$\mu^2 \frac{\partial^2 V}{\partial \phi_b \partial \phi_a} = \delta_{ab} \quad , \tag{3.10}$$

where we used Eqn. (3.6a). This is clearly positive definite and hence shows that the normal vacuum for the free theory corresponds to a local minimum of V.

4. Interactions

We introduce interactions in an approximation defined by a truncated SD equation for the propagator.

$$S^{-1}(p) = \gamma \cdot p - J(p) \; + \tag{4.1}$$

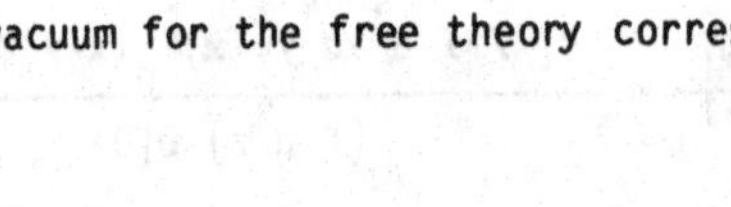

The kernel in the box need not be specified as yet; it could be a photon exchange, a confining propagator, or a chiral invariant meson exchange for example. This approximation is the first term in a systematic approximation developed for example by Cornwall, Jackiw, and Tomboulis.[4] This approximation can be phrased in terms of a variational principle for the generating function W introduced in Eqn. (2.3) as they show. This approximation gives interactions as a modification of Eqn. (3.3) e.g.

$$\frac{\partial^4 W}{\partial J^4} = \qquad (4.2)$$

W contains planar graphs with no vertex corrections.[1]

Again we expand about the chiral limit, $J_n = 0$, $\phi_n = 0$ and again encounter pyramiding difficulty. However, a judicious choice of expansion functions can simplify the problem. Let us choose $F_n(p)$ to be a complete set of eigenfunctions of the Bethe-Salpeter Kernel:

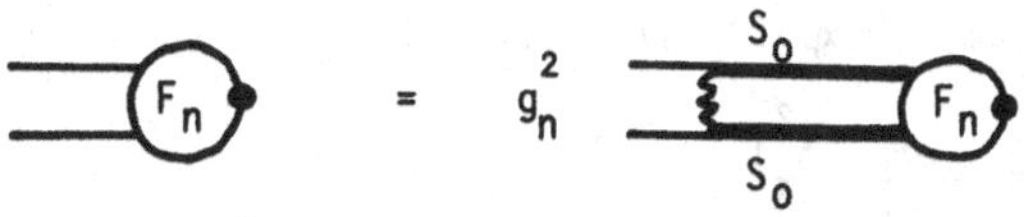

$$= \quad g_n^2 \qquad (4.3)$$

where S_0 is the chiral invariant solution of Eqn. (4.1) with $J(p) = 0$. With this choice of expansion functions, it is very easy to compute the Legendre transform and the result is:

$$V(\phi) = V_{free}(\phi) - \frac{1}{2\mu^2} \sum_n \frac{g^2}{g_n^2} \phi_n^2 \qquad (4.4)$$

The upshot is that the whole effect of the interaction term is to give a quadratic term in the effective potential.

As a check on this expresson we evaluate the Legendre transform relation, Eqn. (3.9), which gives:

$$\sum_n J_n F_n(q) = \gamma \cdot q - \frac{1}{\mu^2} \sum_n \frac{g^2}{g_n^2} \phi_n F_n(q) - S^{-1}(q) \quad . \qquad (4.5)$$

where we have used Eqn. (3.6). This is the S-D equation, Eqn. (4.1) expressed in terms of ϕ_n. Comparing this with Eqn. (3.8) (which is unchanged by interactions), one obtains a relation between the set of J_n sources and the ϕ_n

$$\frac{1}{\mu^2} S_0(k) \left[\sum_a \phi_a F_a(k) \right] S_0(k) =$$

$$\left[K - \sum_n F_n(k) \left(J_n + \frac{1}{\mu^2} \frac{g^2}{g_n^2} \phi_n \right) \right]^{-1} \qquad (4.6)$$

5. Illustration

Let us take the kernel in Eqn. (4.1) to be one photon exchange. In this case, S_0 has the form:

$$S_0^{-1} = \gamma \cdot p\, A(p^2) \qquad (5.1)$$

Further, in order to obtain a discrete spectrum, we cutoff the momentum integrals $\int_\lambda^\Lambda dp$. The analytic solution for A can be given parametrically:

$$p^2 = e^t ,$$

$$A_\pm(\xi) = -\frac{1}{2}(\pm\gamma/2)^{1/2} \exp(\mp\xi^2)/H_\pm(\xi) ,$$

$$t_\pm(\xi) = \frac{1}{2}\int_{\xi_0}^{\xi} d\eta \frac{\exp(\mp\eta^2)}{H_\pm(\eta)} ,$$

$$H_\pm(\xi) = \int_{\eta_0}^{\xi} \exp(\mp\eta^2) d\eta , \qquad (5.2)$$

where $\gamma = \alpha g^2/8\pi^2$ (α is the gauge parameter, g is the gauge coupling constant and the signs $\pm$ refer to the sign of α) and η_0 and ξ_0 are integration constants which can be expressed in terms of λ and Λ. The Bethe-Salpeter equation for F_n in the scalar channel becomes a confluent hypergeometric equation in the variable ξ as shown in Ref. (1).

In the Landau gauge, everything simplifies greatly:

$$S_0^{-1} = \gamma \cdot p$$

$$F_n = (p^2)^{-\frac{1}{2} \pm i\left(\frac{3g^2}{16\pi^2} - \frac{1}{4}\right)^{1/2}} \quad \text{(scalar)} \qquad (5.3)$$

The g_n^2 form a discrete spectrum with the condition $g_n^2 > 4\pi^2/3$.

Let us examine V. The general expression for the second derivatives of V is:

$$\mu^2 \frac{\partial^2 V}{\partial\phi_b \partial\phi_a} = -\frac{g^2}{g_a^2}\delta_{ab} +$$

$$i \frac{1}{\mu^2} \int \frac{d^4k}{(2\pi)^4} \mathrm{Tr}\{F_a[S_0 S^{-1} S_0] F_b [S_0 S^{-1} S_0]\} \quad . \tag{5.4}$$

In the Landau gauge,

$$S^{-1}(p) = \gamma \cdot p - B; \; B \equiv \sum_n \frac{g^2}{g_n^2} \frac{1}{\mu^2} \phi_n F_n(p) \quad , \tag{5.5}$$

giving: (after rotation to Euclidean)

$$\mu^2 \frac{\partial^2 V}{\partial \phi_b \partial \phi_a} = \left(1 - \frac{g^2}{g_a^2}\right)\delta_{ab} - 4 \int \frac{d^4k_E}{(2\pi)^4} F_a F_b \frac{B^2}{k^4} \quad . \tag{5.6}$$

If we examine the normal vacuum, i.e. $B = 0$, then we see that V has a local minimum for g^2 less than the lowest eigenvalue. Beyond that value it becomes a saddle point. (If the channel in question is repulsive then the g_a^2 are negative and no instability occurs.) If we go to a value of g^2 corresponding to an unstable normal vacuum then $B \neq 0$ and the second term in Eqn. (5.6) contributes. Note that this gives a negative definite contribution, and hence further enhances instability.

6. Conclusions

In summary it is noteworthy that the one fermion loop expression for the effective potential, Eqns. (3.7) and (4.4) when properly defined does correspond to the solution of a Schwinger-Dyson equation. The one loop effective potential could very well be useful in phenomological applications.

The instability we discuss here could be symptomatic of a problem that has long been known to plague the truncated Schwinger-Dyson equation, Eqn. (4.1). The solutions have unphysical singularities and these are often diagnostic of instabilities. More work is necessary to draw a firm connection here. A further question should be raised in studies of dynamical symmetry breaking and that is: what guarantees that sacred symmetries are not broken spontaneously - such as Lorentz invariance. In this model for sufficiently large g^2 we fully expect condensates with a non-zero vacuum expectation of a four vector to form. The question is whether there are general principles that inhibit Lorentz breaking.

Finally, work needs to be done in order to incorporate general principles that follow from the renormalizaton group and Ward identities. It would be very useful to exhibit a soluble model which has a stable symmetry breaking minimum.

7. Acknowledgement

This work was supported in part by the U. S. Department of Energy under contract DE-AS05-77ER05490.

8. End Notes

1. Reference 3 gives a more detailed discussion of the work described here. It also contains extensive references.

9. References

[1]. R. Haymaker and J. Perez-Mercader, Phys. Lett. 106B, (1981), 201.

[2]. R. Haymaker, Acta Physica Polonica, B13, (1982), 575.

[3]. R. Haymaker and J. Perez-Mercader, Phys. Rev. D27, (1983), 1353.

[4]. J. Cornwall, R. Jackiw, and E. Tomboulis, Phys. Rev. D10, (1974), 2028.

R. Haymaker and J. Perez-Mercader
Department of Physics and Astronomy
Louisiana State University
Baton Rouge, Louisiana, 70803, USA

INFRARED BEHAVIOR OF QCD

J.S. Ball

Dept. of Physics, University of Utah, Salt Lake City, UT 84112

1. Introduction

In this talk I will describe a series of calculations based on a general non-perturbative continuum approach to the problem of the IR behavior of QCD. This investigation, done in collaboration with F. Zachariasen and M. Baker, is based on using the Schwinger-Dyson eqs. (hereafter S.D. eq.) for the quark and gluon propagators and using the appropriate Ward identities to obtain closed integral equations for the quantities of interest [1]. In particular, we have been able to show that a $1/q^4$ singularity in the gluon propagator produces a consistant solution to the integral equation. When this propagator is used in the integral equation for the quark propagator the pole singularity of a free massless quark is weakened to a simple square root branch point; furthermore we find that a mass-like term (breaking chiral symmetry) which is an entire function, is consistent with the equations, though not required [2]. Finally, we construct an effective classical Lagrangian which is consistent with our form for the gluon propagator and automatically includes the long range quantum fluctuations of QCD [3].

First, let us illustrate the general method for the simpler case of QED. The Dyson eq. for the electron (we will take the electron to be spinless to simplify the algebra) propagator is

$$S^{-1}(P) = S_o^{-1}(P) + e_o^2 \int d^4k \, \Gamma_\mu^o \, D_{\mu\nu}(k) \, S(P+k) \, \Gamma_\nu \tag{1.1}$$

where the bare quantities Γ_μ^o and S_o^{-1} as well as the photon propagator $D_{\mu\nu}$ are assumed known. As was pointed out in an earlier talk by Ken Johnson, this equation might better be called a relation, in that it relates two unknown functions: the electron propagator and the electron photon vertex function.

The Ward identity provides another relation between these quantities

$$q_\mu \Gamma_\mu = S^{-1}(P') - S^{-1}(P) \qquad \text{where} \qquad q = (P' - P) \tag{1.2}$$

which can be "solved" as follows

$$\Gamma_\mu = \frac{S^{-1}(P') - S^{-1}(P)}{P'^2 - P^2} (P + P')_\mu + B(P^2, P'^2) [P\cdot qP'_\mu - P'\cdot qP_\mu] \tag{1.3}$$

where B is a general scalar function. The requirement that Γ_μ be free of kinematic singularities has been used to obtain a unique longitudinal vertex i.e., forms such as

$$\frac{S^{-1}(P') - S^{-1}(P)}{q^2} q_\mu \tag{1.4}$$

are not allowed as they have a kinematic singularity at $q^2 = 0$. Note that as any of the momenta vanish the tensor form multiplying B vanishes, meaning that the first term ($\Gamma^{(\ell)}$) dominates for small q, i.e.

$$\Gamma_\mu \to \Gamma_\mu{}^\ell = \frac{S^{-1}(P') - S^{-1}(P)}{P'^2 - P^2} (P + P')_\mu \quad \text{as } q \to 0.$$

If Γ_μ is replaced by $\Gamma_\mu{}^\ell$ (correct in the IR limit) the Dyson eq. becomes a closed integral eq. for S(P). Choosing the photon propagator

$$D_{\mu\nu} = \frac{\delta_{\mu\nu}}{q^2} Z_3 \tag{1.5}$$

we obtain

$$S^{-1}(P) = S_0^{-1}(P) + e_0{}^2 Z_3 \int \frac{d^4k}{k^2} (P + P')^2 \left\{\frac{1 - S^{-1}(P)\, S(P')}{P'^2 - P^2}\right\} \tag{1.6}$$

which is a linear integral eq. for S(P)

$$S(P) = S_0(P)\,(1 + e^2 \int \frac{d^4k}{k^2} (P + P')^2 \{\frac{S(P') - S(P)}{P'^2 - P^2}\}) \, . \qquad (1.7)$$

The solution for $P^2 \to m^2$ is

$$S(P) \sim (P^2 - m^2)^{-1-\alpha/\pi}$$

which is the correct answer for QED in this gauge.

2. The Gluon Propagator

Now let us turn to the QCD problem. To maximize the similarity to QED, we choose to work with QCD in the axial gauge. In this gauge the Ward identity is the simple generalization (more indices) of that of QED and this gauge has the further advantage of being free of ghosts. The disadvantages of this gauge are as follows: 1) Because of the additional 4-vector n ($n \cdot A = 0$) the tensor forms are more complicated; 2) A second scalar variable $1/\gamma = (n \cdot q)^2/n^2q^2$ exists; and 3) The treatment of $n \cdot k$ denominators in loop integrations requires a special prescription.

2.1 Formulation of Equations

If we now consider the QCD S.D. eq. for a N-point vertex function, we find that the structure is determined by which bare vertex functions are non-zero. The structure is roughly as follows

$$\Gamma_N = \Gamma_N^{(0)} + \int \Gamma_3^{(0)}\, \Delta\Delta\, \overline{\Gamma}_{N+1} + \int \Gamma_4^{(0)}\, \Delta\Delta\Delta\, \overline{\Gamma}_{N+2} \qquad (2.1)$$

where the first integral is a single loop and the second has two loops and the quantities $\overline{\Gamma}_N$ are constructed from Γ_N, Γ_{N-1} ... etc. Thus, because of the bare 4-gluon vertex, closing this equation requires being able to express both Γ_{N+1} and Γ_{N+2} in terms of Γ_N. On the other hand the Ward identity is the same as QED

$$q\, \Gamma_{N+1}^{(\ell)} = \sum \Gamma_N \qquad (2.2)$$

and as before $\Gamma_{N+1} \to \Gamma^{\ell}{}_{N+1}$ when any one of the q's is small. There are two possible routes to closing the S.D. eq. Either we can use the

Ward identity twice to calculate Γ_{N+2} from Γ_N. This approximation can only be good when at least two of the momenta in the vertex are small. This is too restrictive to be of much use in the S.D. eq. for the $\Gamma_2 = \Pi$, the inverse propagator represented by the diagrams in Fig. 1. The momentum of the external lines can be restricted to small values; however, in the four-gluon graph all loop momenta are integrated over all values, making it impossible to restrict two momenta entering this vertex to small values.

The second possibility is to consider only the n-projections of the S.D. eqs. The projection $n \cdot \Gamma_N$ contains only vertices up to Γ_{N+1} as the $\Gamma_4^{(0)}$ is a sum of products of the form $\delta_{\mu1\mu2}\ \delta_{\mu3\mu4}$, hence $n \cdot \Gamma_4^{(0)}$ always has the n index dotted into one of the propagators in the $\overline{\Gamma}_{N+2}$ term $(n \cdot \Delta = 0)$. Thus the terms containing $\overline{\Gamma}_{N+2}$ all vanish from this projection and the Ward identity can be used to close the S.D. eqs. for Γ_N. In the case of Γ_2 the last two diagrams vanish leaving only the first three graphs. However this is only one equation for the two independent scalar functions in Γ_2. To make use of this scheme a further assumption is required to reduce the number of independent tensor forms in Γ_2 or Γ_N so that the n-projections provide enough equations to determine the scalar functions involved. This can be achieved by the ansatz that in the infrared limit the vertex functions are Lorentz invariant (i.e. there is no explicit n dependence in the tensor forms).

With this assumption

$$\Pi_{\mu\nu} = - Z^{-1}\ (q^2\ \delta_{\mu\nu} - q_\mu q_\nu) \tag{2.3}$$

and

$$\Delta_{\mu\nu} = Z\ \Delta_\mu{}^0{}_\nu \tag{2.4}$$

where $\Delta_\mu{}^0{}_\nu$ is the bare gluon propagator and Z is a scalar function of q^2 and γ. If the Ward identity is used to calculate Γ_3 from $\Pi_{\mu\nu}$ i.e. $\Gamma_3 \cong \Gamma^\ell{}_3$ in the small q limit $n \cdot \pi \cdot n$ provides a closed non-linear integral equation for the function Z, which should approach the true Z for small values of q^2. The "solution" to the Ward identity for $\Gamma_3^{(\ell)}$ in terms of Z, leaving out the trivial overall color factor is

$$\Gamma_{\mu_1\mu_2\mu_3}(P_1,P_2,P_3) = \delta_{\mu_1\mu_2}\left(Z^{-1}(P_1)P_{1\mu_3} - Z^{-1}(P_2)\,P_{2\mu_3}\right)$$

$$-\frac{Z^{-1}(P_1) - Z^{-1}(P_2)}{P_1^2 - P_2^2}\,[P_1 \cdot P_2\,\delta_{\mu_1\mu_2} - P_{2\mu_1}P_{1\mu_2}]\,(P_1 - P_2)_{\mu_3} \quad . \quad (2.5)$$

This should be compared to a typical transverse tensor

$$((P_1 \cdot P_2)\,\delta_{\mu_1\mu_2} - P_{1\mu_2}\,P_{2\mu_1})\,[P_1 \cdot P_3\,P_{2\mu_3} - P_2 \cdot P_3 P_{1\mu_3}] \qquad (2.6)$$

which is proportional to $P_1P_2P_3$ and hence vanishes in the limit that any of the momenta go to zero. It should be noted at this point that once Γ_2 has been obtained, the $n \cdot \Gamma_3$ equation can be used to calculate both longitudinal and transverse parts of Γ_3 which can then be used to calculate an improved Γ_2 and check the validity of our assumptions.

To simply the calculation we make the further assumption that Z is independent of γ, which can be checked in the course of solving the resulting integral eq. The final eq. for Z, is

$$-q^2\left(1 - \frac{1}{\gamma}\right)Z^{-1} = -\,q^2\left(1 - \frac{1}{\gamma}\right) + \int dk\; K(k,q,n)\; Z(k) + \int dK\; L(k,q,n)$$

$$\frac{Z(k)\;Z(k')}{Z(q)} \qquad (2.7)$$

where the kernels K and L are known functions of k, q and n.

2.2 Renormalization Proceedure

The first step in this process is to remove all terms proportional to Λ^2 (UV cutoff) so that $\Pi_{\mu\nu}(0) = 0$. Note that if $Z = \Lambda^2/q^2$ is used for all q^2 the integral produces a constant term and this term is also removed in this process. A mass scale is introduced by defining $Z(q^2) = Z(M^2)\;Z_R(q^2)$ which then provides a definition of $g^2(M)$. It should be emphasized at this point that the renormalization proceedure necessary to make our approximate equation finite is not that of the true equation for QCD and it is only in the small q^2 limit that we expect our function Z to have the same behavior as the true Z for QCD.

Renormalization group arguments can then be used to determine the large q^2 behavior of Z:

$$Z \to \frac{1}{\left(1 + \frac{16}{11} b\, g^2(M) \log \frac{q^2}{M^2}\right)^{11/16}} \tag{2.8}$$

where b is the usual $b_{YM} = 11\, N_c/48\pi^2$. The factors of 11/16 are traceable to the missing transverse degrees of freedom.

2.3 Numerical Results

The final integral equation is a non-linear eq. with two non-trivial integrals. Because of numerical difficulties, the integrals can only be calculated accurately for two orders of magnitude variation in q^2; however, the scaling properties of the equation make it possible to cover the full range of q^2 by simply varying $g^2(M)$ and requiring that the Z's in overlapping regions satisfy

$$Z(g^2(M_2), \frac{q^2}{M^2}) = Z(g^2(M_1), \frac{q^2}{M_1^2})\, Z^{-1}\, (g^2(M_1), \frac{M_2^2}{M_1^2}) \quad . \tag{2.9}$$

The general procedure is to choose an input trial function and calculate the resulting output. The parameters of the trial function are then varied until the input and output agree to a few percent over the entire range of q^2. A comparison of the input and output Z^{-1} is shown in Fig. 2. This is done for a fixed value of $\gamma = n^2q^2/(n\cdot q)^2$. After a solution was obtained, the sensitivity to γ was studied by varying it over the range 2-10, see fig. 3. The resulting output Z varied only a few percent, indicating that while Z is actual γ dependent, the dependence is weak enough to justify our assumption. The best fit form (large q^2 behavior is built in) is

$$Z = .16\,\frac{M}{q^2} + \left(\frac{q^2}{q^2 + .75M^2}\right)^{.18} F(q) \tag{2.10}$$

where $g^2(M) = \frac{4\pi}{N_c}$ and $F(q) = \left\{1 + \frac{16}{11} b\, g^2(M) \log\,(.2 + .8\, q^2/M^2)\right\}^{-11/16}$.

Thus we have shown that a consistent solution exist in which Z behaves like $1/q^2$ in the IR limit where it should agree with the true Z for QCD.

2.4 Analytic Results in IR Limit

Our numerical results are rather insensitive to the power behavior of the first non-leading term at small q^2. A check of this result plus a test of our numerical method can be carried out as follows: we separate out the pole behavior of Z:

$$Z = \frac{C}{q^2} + Z_1 \qquad (2.11)$$

and then linearize the equation for Z_1, which should be valid at small q^2. If we now introduce a spectral representation for Z_1

$$Z_1 = \int \frac{d\mu^2 \, \rho(\mu^2)}{q^2 + \mu^2} \qquad (2.12)$$

we can, in the kinematic limit $n \cdot q = 0$, perform all integrations to obtain the renormalized S.D. eq.:

$$Z^{-1} = 1 + \frac{g\,(M)}{8\pi^2} \int_0^\infty \rho \, \frac{(q\,x)dx}{x} \left\{K(x) - K\left(\frac{q}{M^2}\,x\right)\right\} \qquad (2.13)$$

where $x = \frac{\mu^2}{q^2}$ and $K(x)$ is a known function.

If we look at the small q^2 behavior of Z^{-1} we find

$$\frac{\frac{q^2}{C}}{1 + \frac{q^2}{C} Z_1} \approx \frac{q^2}{C} - \frac{q^4}{C^2} Z_1 \,. \qquad (2.14)$$

Thus, if $Z_1 = (q^2)^\nu$, then it must not appear as a $(q^2)^\nu$ on the right hand side of the D.S. eq. If we let $\rho = (\mu^2)^\nu$, consistency requires that

$$q^\nu \int \frac{dx}{x} x^\nu K(x) = 0 \; . \tag{2.15}$$

This integral can be done analytically and we find $\nu = .1737$ which is in good agreement with our previous result and shows that a power law correction to the pole term is in fact the correct behavior.

3. Quark Propagator

It is commonly believed that there is some causal relationship between confinement and chiral-symmetry breakdown [4]. Since we believe that the $1/q^4$ singularity which we have found to be the self consistent IR behavior for the gluon propagator is an indication of confinement, it may be that the existence of this singular behavior implies directly some quark mass generation.

It is clear that the gluon propagator obtained in the previous section provides necessary input to the S.D. eq. for the quark propagator. If the Ward identity for the quark-gluon vertex is used one obtains a closed linear integral eq. for the scalar functions in the quark propagator.

Let us begin this proceedure by defining these functions in axial gauge for the unrenormalized quark propagator as follows:

$$S(P) = (\not{P}F + G) + (\not{P}H + I)\not{n} \quad . \tag{3.1}$$

For the free propagator $S_o(P)$ (massless quarks) $F = 1/p^2$ while G, H, and I are all zero.

The quark propagator satisfies the following S.D. eq.:

$$S^{-1}(P) = S_o^{-1}(P) + g_o{}^2 \int dk \; \gamma_\mu \; \Delta_{\mu\nu} \; (k) \; S(P - k) \; \Gamma_{\nu(p - k, \; P)} \tag{3.2}$$

where we will take Δ to be the infrared singular gluon propagator obtained in the previous section

$$\Delta_{\mu\nu}(k) = - Z(M) \frac{AM^2}{k^4} (k^2 \Delta_{0\mu\nu}) \tag{3.3}$$

and where Γ_ν is the quark gluon vertex function. We will again use the Ward identity to calculate the longitudinal vertex which dominates in the infrared.

$$\Gamma_\mu^{(\ell)}(P - k, P) k_\mu = S^{-1}(P) - S^{-1} (p-k) . \tag{3.4}$$

After multiplication of eq. (3.3) on the right by S(P) and on the left by $S_0(P)$ we obtain the following:

$$S_0(P) = S(P) + g_0^2 S_0(P) \int dk\, \gamma_\mu \Delta_{\mu\nu} S(P') \Gamma_\nu S(P) . \tag{3.5}$$

The projections of this equation become a set of linear integral equations for the scalar functions F, G, H, and I when the following "solution" to the Ward identity is used:

$$S(P') \Gamma_\nu^{(\ell)} (P',P) S(P) = - \frac{1}{2} \{(F(P) + F(p'))\gamma_\nu + (\frac{F(P) - F(P')}{p^2 - p'^2})$$

$$(2\not{P}' \gamma_\nu \not{P} + (P^2 + P'^2) \gamma_\nu) + 2 (\frac{G(P) - G(P')}{p^2 - p'^2}) (\gamma_\nu \not{P} + \not{P}'\gamma_\nu)\}$$

$$- (\text{same with } F \to H\ G \to I) \not{M} . \tag{3.6}$$

Here we have assumed that the scalar functions are independent of the gauge variable $(n\cdot p)^2/n^2p^2$. Before discussing a detailed solution of this equation let us examine some of the general features. Note that the functions F and I are always multiplied by an odd number of γ-matrice while G and H have even numbers of γ's. Therefore after taking traces to project out the scalar equations we will have 2 linear integral equations which contain F and I and two which contain G and H but no coupling between these equations. Furthermore, since S_0 has both G and H zero there will be no inhomogeneous terms in that pair of equations. This means that G and H identically equal to zero will always be a possible solution! This decoupling of the chiral invariant terms and the mass type terms is quite general and is not altered by including transverse parts of the vertex or by the form of the gluon propagator Thus, while solutions that break chiral symmetry exist, there are always solutions to these equations which preserve the symmetry.

The analysis of these equations is greatly simplified if we constrain the external momentum so that $n \cdot p = 0$. Since we have already assumed that F and G are independent of $n \cdot p$ nothing of importance should be lost by this choice. In this limit the algebra is greatly simplified and the functions H and I can be dropped. The resulting equations are

$$1 = p^2 F - g_0^2 \int dk\, \Delta_{\mu\nu}(k) \{\frac{1}{2} (F + F')\, \delta_{\mu\nu} + \frac{F - F'}{p^2 - p'^2} (\frac{1}{2} k^2\, \delta_{\mu\nu} + P_\mu P_\nu' + P_\mu' P_\nu)\} \tag{3.7}$$

and

$$0 = p^2\, G - g_0^2 \int dk\, \Delta_{\mu\nu}(k)\, (\frac{G - G'}{p^2 - p'^2})\, [(p^2 - p \cdot p')\, \delta_{\mu\nu} + P_\mu P_\nu' + P_\nu P_\mu'] \tag{3.8}$$

which, with no approximations, can be written as the following differential equations for F and G:

$$1 = p^2\, F(p^2) + \beta M^2\, p^2\, \frac{\partial F(p^2)}{\partial p^2} - \frac{1}{2}\, \beta M^2 F(p^2) \tag{3.9}$$

and

$$0 = p^2\, \beta M^2\, \frac{\partial^2 G(p^2)}{(\partial p^2)^2} + (p^2 + 2\beta M^2)\, \frac{\partial G(p^2)}{\partial p^2} + (2 - \frac{\beta M^2}{2})\, G(p^2). \tag{3.10}$$

Here $\beta = AN_F\, g_A^2(M)/4\pi^2$, where $g_A^2(M)$ is the axial-gauge coupling constant renormalized at M and N_F is the quark Casimir eigenvalue. The solution for F is

$$F_R(p^2) = - (\beta M^2)^{-1}\, \phi(1, 1/2; - p^2/\beta M^2) \tag{3.11}$$

where ϕ is a confluent hypergeometric function which behaves like $1/p^2$ as $p^2 \to \infty$ consistent with the requirements of asymptotic freedom and for the IR limit

$$F_R(p^2) \to \text{constant} + \sqrt{p^2} \quad . \tag{3.12}$$

It should be mentioned that with different boundary conditions, there is a solution for F which is an entire function of p^2; however such a function cannot be consistant with the usual asymptotic freedom limit in the UV.

The only solution to the differential equation for G which is compatible with the integral equation is

$$G(p^2) = C\,\Phi\,(2,\, 5/2;\, -\, p^2/\beta M^2) \tag{3.12}$$

which is an entire function of p^2. Clearly C = 0 yields the chiral symmetric solution.

The results of using our singular gluon propagator to calculate the quark propagator can be summarized as follows: The singularity of a bare quark propagator $S = \not{p}/p^2$ is "softened" to $S = \not{p}\,(\text{const.} + \sqrt{p^2})$ i.e. a pole is changed into a squareroot branch point. Chiral symmetry can be broken but it is not required that it be broken. The quark propagator cannot be an entire function as this is not compatible with asymptotic freedom.

4. An Effective Action for Long Range QCD

Our calculations of the quark and gluon propagators, which are gauge variant quantities, have been done in axial gauge because of the simplicity of the Ward identities and the absence of ghosts, in that gauge simplified the calculation. While the behavior of these gauge variant quantities are of interest and the location of their singularities are in fact gauge invariant (masses and thresholds), we would like to use these quantities to calculate physical observables which will of course be gauge invariant. These steps are exactly those used in QED perturbation calculations; however, we cannot use perturbation methods for constructing observables and we cannot use perturbation theory to lowest order to construct "gauge invariant propagators". The best approach at this point seems to be using our propagator and the Ward identities, which are an expression of the gauge invariance of the theory, to construct an effective action which includes the long range (IR) quantum fluctuations of QCD. In the standard field theory formalism a classical external color current source $J_\mu^a(x)$ is introduced in constructing the generating function. In this case the vacuum expectation value of the field <A> is a functional of J_μ^a. We can however also invert this relationship and think of J as a functional of <A>. In this case we could expand J_μ in a Taylor series in <A> as follows:

$$J_\mu^a(x) = \int \left. \frac{\delta J_\mu^a(x)}{\delta \langle A_\nu^b(y) \rangle} \right|_{\langle A \rangle = 0} \langle A_\nu^b \rangle \, d^4y$$

$$+ \frac{1}{2} \iint \left. \frac{\delta^2 J_\mu^a(x)}{\delta \langle A_\nu^b(y) \, \delta \langle A_\lambda^c \rangle} \right|_{\langle A \rangle = 0} \langle A_\nu^b(y) \rangle \langle A_\lambda^c(z) \rangle d^4y \, d^4z$$

$$+ \ldots \tag{4.1}$$

These expansion coefficients are simply the proper n-legged vertex functions. For example

$$\left. \frac{\delta J_\mu^a(x)}{\delta \langle A_\nu^b(y) \rangle} \right|_{\langle A \rangle = 0} \qquad \Pi_{\mu\,\nu}^{\,a\,b}(x,y) \tag{4.2}$$

$$\left. \frac{\delta^2 J_\mu^a(x)}{\delta \langle A_\nu^b(y) \, \delta \langle A_\lambda^c(z) \rangle} \right|_{\langle A \rangle = 0} = \Gamma_{\mu\,\nu\,\lambda}^{(3)}(x,y,z) \tag{4.3}$$

If pure longitudinal terms dominate the Γ's we can calculate them from $\Pi_{\mu\nu}$ and the appropriate Ward identities. Keeping only the IR singular behavior of $\Pi_{\mu\nu} \sim q^4$ we can calculate all Γ's up to N = 6 and $\Gamma_N = 0$ for N > 6. The reason for this is each higher Γ has one less power of q therefore $\Gamma_2 \sim q^4$ means $\Gamma_3 \sim q^3$ ... and $\Gamma_6 \sim \delta_{\mu\nu}$'s just as the Γ_4 was for the bare quantities.

If we consider the weak field limit of eq. (4.1) where $\Gamma_{\mu\nu} = Z^{-1}(q)$ $(q^2\,\delta_{\mu\nu} - q_\mu\,q_\nu)$ and keep only the first term in the expansion for J_μ^a, we obtain

$$J_\mu{}^a(q) = i\, q_\nu\, Z^{-1}(q)\, \langle F_{\nu\mu}{}^a(q)\rangle \tag{4.4}$$

where we would identify $Z^{-1}(q)$ as the dielectric constant of a medium

$$\varepsilon(q) = \mu^{-1}(q) = Z^{-1}(q) \sim q^2 \tag{4.5}$$

and the corresponding Maxwell eq.

$$J_\mu{}^a(x) = \frac{1}{M^2}\ \partial_\nu\, \partial^2 \langle F_\mu{}^a{}_\nu\rangle\ . \tag{4.6}$$

Note that the solution for a point charge is

$\vec{E} = Q\vec{r}/r$ corresponding to a linear potential.

If we keep all Γ's we obtain the following equation:

$$J_\mu(x) = \frac{1}{gM^2}\, D_\nu D^2 \langle F_{\mu\nu}(x)\rangle - \frac{i}{2gM^2}\, [D_\mu\, F_{\lambda\sigma}(x),\ F_{\lambda\sigma}(x)] \tag{4.7}$$

where

$$D_\mu = \partial_\mu - i\, [\langle A_\mu\rangle\ ,]\ .$$

These field equations are a consequence of the following effective action:

$$L_{eff} = -\frac{1}{2} \langle F_{\mu\nu}{}^a(x)\rangle \frac{D^2}{g^2M^2} \langle F_{\mu\nu}{}^a\rangle\ . \tag{4.8}$$

These equations describe an inhomogeneous anisotropic classical medium with a field dependent dielectric tensor given by the following operator:

$$\varepsilon_{\mu\lambda} = \frac{1}{M^2 g}\, (\delta_{\mu\lambda}\, D^2 + D_\mu\, D_\lambda)\ . \tag{4.9}$$

The dielectric constant ε and the magnetic permeability μ are given by

$$\varepsilon = \mu^{-1} = \frac{D^2}{M^2 g} . \tag{4.10}$$

The Hamiltonian in this medium is

$$H = -\frac{1}{2}\vec{E}\cdot\frac{\vec{D}^2}{M^2g^2}\vec{E} - \frac{1}{2}\vec{B}\cdot\frac{\vec{D}^2}{M^2g^2}\vec{B} + \frac{3}{2}\vec{E}\cdot\frac{D_o^2}{M^2g^2}\vec{E} - \frac{1}{2}\vec{B}\cdot\frac{D_o^2}{M^2g^2}\vec{B}. \tag{4.11}$$

To gain an understanding of this medium consider the solution exterior to a color electric or magnetic charge. Since we expect our results to be valid only at long range i.e. $r > 1/M$ we will consider the fields given on the surface of a sphere of radius $R_o = 1/M$ and look for spherically symmetric Abelian solutions outside of this sphere.

First consider $\vec{B} = 0$ and a color electric change (q) inside the sphere:

$$\vec{D} = \varepsilon\,\vec{E} \qquad\qquad \varepsilon = -\nabla^2 . \tag{4.12}$$

Outside the sphere we have

$$\nabla\cdot\vec{D} = 0 \qquad \text{hence} \qquad \vec{D} = \frac{q\,\vec{r}}{4\pi r^3} \tag{4.13}$$

and

$$E = \frac{q\,\vec{r}}{8\pi r} \tag{4.14}$$

$$H = \int_{r>R_o} d^3r\,\vec{D}\cdot\vec{E} = \frac{q}{16\pi}\int_{1/M} dr \to \infty . \tag{4.15}$$

Therefore any spherically symmetric color electric flux on the sphere produces an infinite energy in the exterior region; hence we must have color singlet electric sources.

Now consider a color magnetic source (q_m) inside with $E = \vec{0}$.

$$B = \frac{q_m \vec{r}}{4\pi r^3} \qquad \vec{H} = -\nabla^2 \vec{B} = 0 \tag{4.16}$$

where the resulting energy is

$$H = \int B \cdot H = 0 \,. \tag{4.17}$$

Thus the vacuum produced by the QCD IR fluctuations is that of a perfect paramagnet, producing a magnetization equal and opposite to the B field.

The only non-Abelian static solution which we have found is for a uniform color electric line charge. The equations for the potentials in this case are

$$(-\nabla^2 + A^2)^2 \phi + A\phi(\nabla^2 + \phi^2)A + 3(\nabla A)^2\phi - 3A^2 \phi^3 = M^2\rho \tag{4.18}$$

and

$$(\nabla^2 + \phi^2)^2 A + A\phi(-\nabla^2 + A^2)\phi - 3(\nabla\phi)^2 A - 3\phi^2 A^3 = 0 \tag{4.19}$$

where A is the magnitude of $\vec{A}$ in z direction (parallel to line charge). These equations have been solved numerically and have the following general properties:

The behavior at large r is

$$\nabla^2 A = 0 \qquad A = \alpha + \beta \ln r$$

and $(-\nabla^2 + A^2)^2\phi = 0$ which causes ϕ to decrease exponentially

$$\phi \simeq e^{-|A|r} \,.$$

Thus A^2 acts like a mass term to screen the line charge. The resulting energy per unit length is finite in contrast to the Abelian Coulomb solution which has $D \sim 1/r$ and $E \sim r \ln r$ and has infinite energy per unit length.

At this point we have not been able to solve the two point charge (classical) problem nor has it been solved in the simpler case of classical Yang-Mills. In addition to this problem, we are trying to construct a semi-classical model which will be able to treat quantum color sources so that the screening behavior for these sources can be studied.

References

[1] M. Baker, J.S. Ball and F. Zachariasen, Nucl. Phys. B 186 (1981) 531, M. Baker. J.S. Ball and F. Zachariasen, Nucl. Phys. B186 (1981) 560.

[2] J.S. Ball and F. Zachariasen, Phys. Letters 106B (1981) 133.

[3] M. Baker and F. Zachariasen, Phys. Letters 108B (1982) 206.

[4] See for example J.M. Cornwall, Phys. Rev. D22 (1980) 1452 and J.M. Cornwall, U.C.L.A. Preprint 81/TEP/8 (Feb. 1981).

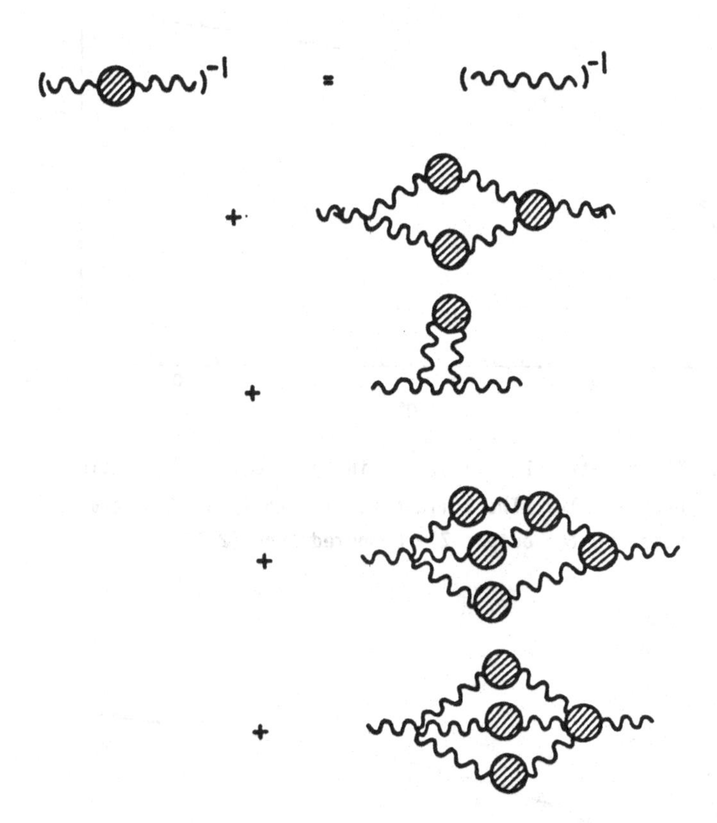

Figure 1. Pictorial representation of the Schwinger-Dyson equation relating $\Gamma^{(2)}$, $\Gamma^{(3)}$, and $\Gamma^{(4)}$.

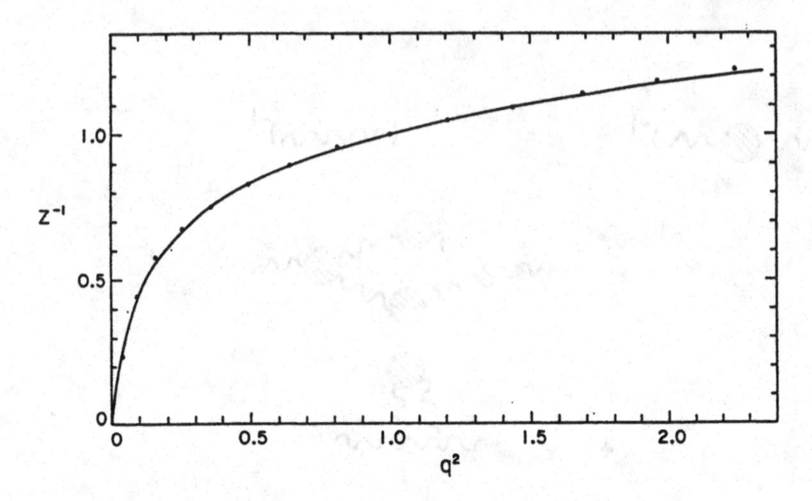

Figure 2. The solution $Z^{-1}(q)$ of the integral eq. (2.7) obtained numerically. The solid curve is the input Z^{-1}, the dots are the output Z^{-1} computed from (2.7).

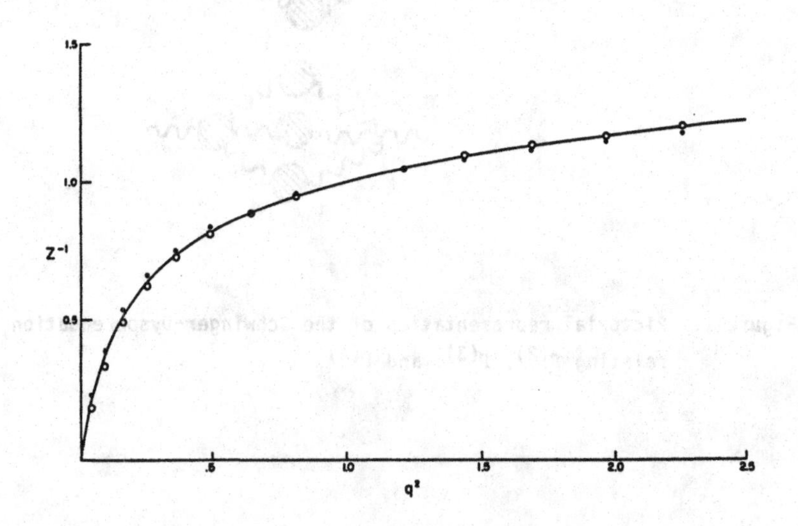

Figure 3. The variation of Z^{-1} computed from eq. (2.7) for various values of γ. The solid curve is the input Z^{-1} optimized for $\gamma = 4$ as shown in Figure 2. The output Z^{-1} for $\gamma = 10$ are the black circles, while those for $\gamma = 2$ are the open circles.

INFRA-RED BEHAVIOR OF THE GLUON PROPAGATOR IN QCD*

Porter W. Johnson
Illinois Institute of Technology and
University of Groningen

The status of recent attempts to determine the infra-red behavior of the gluon propagator in quarkless QCD through solution of truncated Dyson-Schwinger equations is surveyed. The situation in time-like axial gauges, as well as covariant gauges, is discussed.

The gluon propagator, as well as the corresponding vacuum polarization tensor, is gauge dependent in QCD. The recent flurry of interest in its infra-red (IR) behavior in non-singular gauges has been sparked by hope of obtaining a deeper understanding of the IR character of QCD. The gauge-dependent propagator represents a "Born-approximation" to the gauge-independent Wilson loop potential, and it is reasonable (but not mandatory!) to expect its IR behavior to bear a direct relation to confinement. Indeed, G. West has shown that a p^{-4} singularity in the gluon propagator at small momentum p (in <u>any</u> gauge) requires an "area law" for Wilson loop potential [1].

As a manifestation of gluon self-coupling, the Dyson-Schwinger (DS) equation for the gluon propagator is non-linear, and it is open-ended in that it requires knowledge of the triple-and quadruple-gluon vertex functions; Γ_3 and Γ_4, respectively. The Slavnov-Taylor (ST) identities serve to specify the "longitudinal" parts of these vertex functions; whereas the "transverse" parts, which are expected to have kinematical zeros at zero momentum, should be less important in the IR. By making a truncation of the DS system with approximations based upon the ST identities, one obtains a non-linear integral equation for the gluon propagator. This non-linearity opens the door to the possibility of suppression of low-frequency modes of the gluon field, as a manifestation of confinement. I shall describe various truncation schemes that have been proposed in quarkless QCD, beginning with axial gauges and followed by covariant gauges.

The color vector potential is taken orthogonal to a fixed 4-vector n in an axial gauge, so that the gluon propagator $D_{\mu\nu}(n,p)$ (color indices suppressed) is orthogonal to n^μ and n^ν. The free propagator is

$$D^o_{\mu\nu}(n,p) = [g_{\mu\nu} - \frac{n_\mu p_\nu + p_\mu n_\nu}{n\cdot p} + \frac{n^2 p_\mu p_\mu]}{(n.p)^2} \frac{1}{p^2} \quad . \tag{1}$$

The full propagator may be written in terms of the scalar function Z as

$$D_{\mu\nu}(n,p) = Z(n,p)\, D^o_{\mu\nu}(n,p) \tag{2}$$

where a second tensor structure, which is not usually felt to be important in the IR, has been dropped. The DS equations and the ST identities do not involve coupling to "ghost" fields in axial gauges. The DS equations for the gluon propagator may be represented schematically as

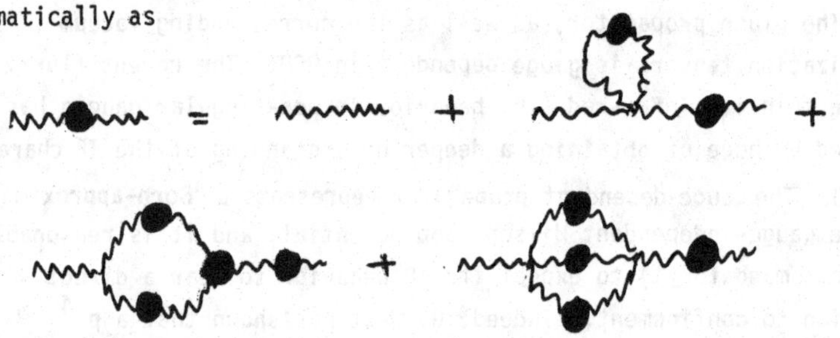

(3)

The solid circles refer to "full" propagators and vertex functions. The equation may be expressed formally as

$$D_{\mu\nu} = D^o_{\mu\nu} + D^o_{\mu\lambda}\, \Pi^{\lambda\rho}\, D_{\rho\nu} \quad , \tag{4}$$

where $\Pi^{\lambda\rho}$ is the vacuum polarization tensor for the gluon field.

One may project out the two-loop diagram in (3) by using the fact that its contribution to the vacuum polarization tensor is orthogonal to the tensor $n_\lambda n_\rho$. The projected DS system has the form [2]

$$p^2 n^2 \left[\frac{1}{Z(p,n)} - 1 \right] (1-\gamma) = n_\mu\, \Pi^{\mu\nu}(p,n)\, n_\nu \quad , \tag{5}$$

where $\gamma = (p.n)^2/(n^2p^2)$ is a gauge parameter. After the form of the 3-gluon vertex Γ_3 is specified, one obtains a closed equation for the scalar function Z(p,n). The triple vertex $\Gamma_3{}^{\mu\nu\lambda}(p,q,r)$, which couples gluon lines of (momentum,spin) indices (p,μ), (q,ν), and (r,λ), must have Bose symmetry. In addition, it should satisfy the ST identities:

$$p_\lambda \, D_{\mu\mu'}(q) \, D_{\nu\nu'}(r) \, \Gamma_3{}^{\lambda\mu\nu}(p,q,r) = D_{\mu'\nu'}(r) - D_{\mu'\nu'}(q) \quad , \tag{6}$$

along with cyclic permutations.

The approach of Baker et al [2,3] has been to make an algebraic Ansatz for Γ_3 which is consistent with Bose symmetry and the ST identity, so that the projected DS equation becomes a non-linear integral equation for $Z(p,n)$. As has been described at this conference by J. Ball, the following conclusions are drawn from a numerical study of that equation: (1) Z is of order p^{-2} in the IR, as an indication of confinement, and (2) Z depends only slightly, if at all, on the gauge-fixing parameter γ. These numerical conclusions have not yet been substantiated by analytic work.

A second approach, advocated by Delbourgo [4], involves a spectral Ansatz for Γ_3, and results in a linear equation for $Z(p,n)$. Its solution is independent of the gauge parameter γ, and $Z(p^2)$ is of order $(\ell n\, p^2)^{-1}$ in the IR. However, the ST identity is not satisfied in this approach, as is evidenced by the fact that a linear equation is obtained for Z.

The approach of our Groningen-IIT collaboration [5] will be described in some detail. Like Delbourgo, we make special Ansätze for Γ_3 and Z. Our Ansatz for Γ_3 is not Bose symmetric, and it satisfies the ST identity (6) only in the external leg. However, the vacuum polarization is transverse in our approach. The Lehmann spectral representation for the gluon propagator corresponds to the formula (Euclidean momenta).

$$Z(p,n) \quad = \quad p^2 \int_0^\infty dt \;\; \rho(t,\gamma(p)) \; \frac{1}{p^2+t} \quad . \tag{7}$$

We analyze the IR behavior and γ-dependence of the solution(s) of the truncated DS equation (5). To that end, we set $\gamma = 0$ in (5) and (temporarily) ignore any γ-dependence in the spectral density ρ. We also assume that the IR behavior of Z is dominated by the term $- a\delta'(t)$ in ρ, and integrate eq.(7) by parts to obtain

$$Z(p,n) \quad = \quad \frac{a}{p^2} + \quad p^2 \int_0^\infty dt \; \frac{\bar{\rho}\,(t)}{(p^2+t)^3} \quad , \tag{8}$$

where the self-consistently determined function $\bar{\rho}(t)$ is to vanish at small t. A similar modification is made in our Ansatz for Γ_3, so that the loop integrals in the truncated DS equation (5) are convergent. Our regularization procedures can thus be expressed entirely in terms of divergences of spectral integrals. Eq.(5) has the form

$$p^2\,[Z^{-1}(p)-1] \;=\; aA + \pi^2 \int_0^\infty dt\,\bar{\rho}\,(t)[-\frac{3}{t} + \frac{11p^2}{3t^2} - \frac{23}{6}\frac{p^4}{t^3} + \ldots]\ , \tag{9}$$

where the constant A arises from a divergent, momentum-independent integral. We regularize (9) by matching the terms of order p° and p^2, in correspondence to wave function and mass renormalization. The p^4 terms in (9) then match automatically, and it appears quite likely that there is a "confining" solution of the regularized equation with the IR asymptote $Z \sim ap^{-2}$.

Does the γ-independent distribution $\bar{\rho}\,(t)$ satisfy (5) for all γ? The left side of (5) is proportional to $1-\gamma$ in such a case, whereas the right side has intricate γ-dependence arising from tensor structures in D (eq.(1).) and Γ_3. An example of this γ-dependence is seen in the following principal value integral over loop momentum:

$$P\int \frac{d^4q}{p.n+q.n}\,\frac{1}{(q^2+M^2)^2} \;=\; 2\pi^2(p.n)\int_0^1 dx\,[M^2n^2 + (p.n)^2(1-x^2)]\ . \tag{10}$$

There can be no γ-independent solutions of the truncated DS system (5) in our approach, because the γ-dependence does not drop out. We note in passing that intricate γ-dependence (e.g., integrals of type (10)) is also a feature of the scheme of Baker et al, and it would be truly remarkable (but not necessarily impossible) for a γ-independent solution Z(p) to occur there.

The truncation of DS equations in covariant gauges will now be reviewed. One circumvents explicit gauge-dependence in the propagator, but the following problems do arise:

1. In contrast to axial gauges, the DS equation and ST identities involve coupling to ghost fields.
2. There is no natural way to separate out the two-loop diagrams in the DS equation.
3. The ghost ST identities are "non-recursive", in that they relate longitudinal parts of n-point functions to, say, (n+1)-point functions. [6]

These difficulties considerably obscure our understanding of covariant gauges.

Mandelstam [7] recently described a truncation scheme in which two loop diagrams and ghost couplings are dropped, with one full propagator and full vertex replaced by bare values in the remaining one-loop term. The scheme is qualitatively consistent with the ST identity in that one extra propagator pole and vertex zero are canceled. The gluon propagator in Landau gauge is

$$D_{\mu\nu}(p) = F(p)\,(g_{\mu\nu} - \frac{p_\mu p_\nu)}{p^2} \quad . \tag{11}$$

Mandelstam suggests that F(p) approaches p^{-4} in the IR, as an indication of confinement.

Our group has analyzed Mandelstam's scheme rather thoroughly [8]. We find that $F(p^2)$ does have the IR asymptotic behavior p^{-4}. However, it also has a number of branch points at complex p^2, which lie in the left-half p^2 plane and appear to accumulate at $p^2 = 0$. These branch points invalidate the Wick rotation made in going to Euclidean momenta, and they cast doubt upon any physical interpretation of the gluon propagator.

It is our feeling that the unphysical results are an artifact of Mandelstam's approximation scheme. Complex branch points arise in similar truncation schemes in QED [8], and it is actually rather important to make physically consistent truncations. We are currently investigating an approach which is similar to Mandelstam's, with the important difference that our truncated vacuum polarization tensor is a transverse object. Work is in progress to explore the IR character of the gluon propagator in our approach.

* Work supported by National Science Foundation grant PHY-81-06908.

[1] G. West, Phys. Lett. B115, 468 (1982).

[2] For a recent summary see M. Baker, "The Running Coupling Constant Yang-Mills Theory", Lectures at Zakopane Summer School, May 1980, CERN TH-2905.

[3] M. Baker et al, Nucl. Phys. B186, 531, 560 (1981).

[4] R. Delbourgo, J. Phys. B186, 531, 560 (1981).

[5] D. Atkinson et al, "IR Behaviour of Gluon Propagator in Axial Gauges", Groningen preprint 1982.

[6] S.K. Kim and M. Baker, Nucl Phys B164, 152 (1980).
[7] S. Mandelstam, Phys. Rev. D20, 3223 (1979).
[8] D. Atkinson et al, J. Math. Phys. 22, 2074 (1981); 23, 1917 (1982).

INFRARED PROPERTIES OF UNUSUAL INITIAL STATE INTERACTIONS

Charles A. Nelson
Department of Physics, SUNY, Binghamton, New York 13901

ABSTRACT

The related roles of gauge invariance and of initial coherent/degenerate states in the factorization of QCD processes into short-distance and long-distance parts are discussed. For $\pi\pi\to\mu^+\mu^- + X$ it is shown that although the systematic summation over all possible spectator interactions removes the outside soft-IR divergences in the non-overlapping ladder Glauber diagrams, unphysical inside soft-IR divergences persist. So, the on-shell Glauber region is not a gauge invariant concept which can be physically isolated from radiative corrections which non-trivially involve other diagrammatic regions. It is conjectured that in QCD non-factorization is inconsistent with local gauge invariance.

The free-field limit of QCD is of course not invariant under local color phase transformations. This fact, and the property of asymptotic freedom, motivate the systematic introduction of initial and final coherent/degenerate states (CS/DS) as a dynamical basis for properly separating the long-distance asymptotic dynamics from the standard short-distance dynamics in perturbative QCD. This CS/DS approach is supported by the demonstrations that the non-cancellation of non-factorizing, soft-IR divergences[1] in purely active interactions in $d\sigma/dQ^2$ for the Drell-Yan process (at the two-loop level in the non-leading power of Q^2) can be removed by using coherent initial quark-gluon states[2] or by taking into account initial quark/gluon degeneracy[3]. Although it has not been necessary here to assume an adhoc non-perturbative QCD induced cutoff, a scale ω has actually been implicitly introduced to characterize the initial soft-gluon region. Physically, for the purely active interactions this quantity ω has the role of a resolution or "coherence scale" for specifying separately each initial active quark or hard gluon leg and consequently the "coherence length" ω^{-1} should be of the order of Λ^{-1} so as not to improperly extend (or restrict) the perturbative dynamics versus the non-perturbative domain. Analogy with surface energy considerations for Type II superconducting systems supports the conclusion that it is necessary that ω^{-1} is less than the effective, hard gluonic penetration length; such an analogy is suggested, for instance, by Fukuda's study[4] of gluon condensation. For $\omega \sim O(\Lambda)$, Van der Waals forces are of course absent between sufficiently separated hadrons. For composite objects, such as hadrons and perhaps more interestingly, jets, it may turn out that additional resolution lengths are required and that these scales can be predicted, e.g. via lattice calculations. This CS/DS approach assumes that collinear infrared divergences are handled by the standard factorization procedure and assumes, implicitly, an appropriate asymptotic infrared renormalization. Besides low-order analyses, the KLN theorem has been generalized to QCD for certain cross-sections in specific processes[5] and there are some arguments[6,7] to arbitrary orders in g for the cancellation of soft-IR divergences in QCD with asymptotic coherent states.

Single-cut Kinoshita diagrams are a useful tool for exorcising the diseases of candidates for non-factorization effects in QCD: For instance, Frenkel-Taylor-and-collaborators[3] considered the $g^4 C_F C_{YM}$ contribution to the color and spin-averaged cross section for the Drell-Yan process $h_1 h_2 \to \mu^+ \mu^- X$ and found to next to leading order in the dimensional regularization parameter, i.e. to $O(1/\varepsilon)$, that Fig. 1 holds unlike in QED. This non-Abelian IR non-cancellation is

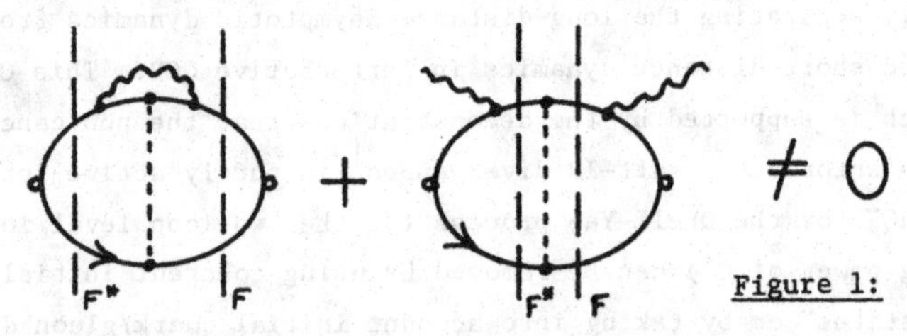

Figure 1:

avoided if the initial "$q\bar{q}$ + gluon DS" are included because of the "Kinoshita diagram identity" to $O(1/\varepsilon)$ shown in Fig. 2. This calculation was performed in the Coulomb gauge in the rest frame of the (lower) anti-quark line. Each cut of such a Kinoshita diagram denotes a Feynman

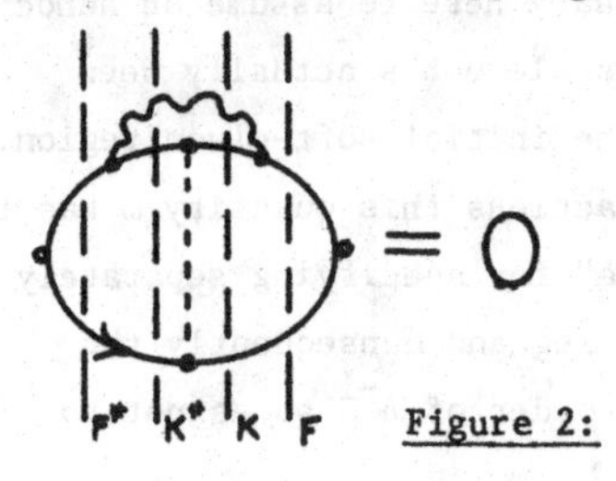

Figure 2:

diagram contribution to the cross section. Disconnected single-cut diagrams do not need to be considered for they are eliminated by S matrix normalization.

On the other hand, to make a gauge invariant statement, it is necessary to sum over sets of such Kinoshita diagrams. In fact, for the above example although it can be shown that the $O(1/\varepsilon)$ coefficient separately vanishes for each single-cut diagram assuming inital DS, without DS there is already a considerable cancellation since the sum of all Bloch-Nordsieck (BN) contributions has a higher twist coefficient.

At one-loop level, if gauge invariance is not considered, e.g. by considering $q\bar{q}$ annihilation into an external colored current, (i) for massless quarks IR cancellation does not take place in the BN cross section,[8,9] and (ii) there are finite contributions to the cross section from the initial, soft DS. When the larger class of diagrams are included as dictated by local gauge invariance,[9] at one-loop level the BN summation is sufficient and <u>the large second-order π^2-type term</u> in $d\sigma/dQ^2$ from $q\bar{q} \to \gamma X$ <u>is not reduced</u> since the effects of the soft DS cancel out among themselves leaving no finite contribution to the cross section.

Next, we will use these single-cut diagrams to briefly review an analysis of the infrared properties of two-loop, <u>initial state spectator interactions</u> in the Drell-Yan process. Specifically, we will use single-cut diagrams to test a Glauber, multiple scattering <u>candidate</u>[10] <u>for leading-twist non-factorization effects</u> in $\pi_1\pi_2 \to \mu^+\mu^- X$:[7,13] This candidate has been discussed by a number of authors, see ref. 11, 12. In light-cone time-ordered perturbation theory a typical single-cut contribution in the $\pi_1(1,\vec{r})$, $\pi_2(1,\vec{0})$ frame is Fig. 3 where the solid arrow denotes a sum over transverse (non-solid arrow) and instantaneous gluon exchanges. In the "dynamical" region for $m_\mu = (z, \vec{m}^2/z, \vec{m})$ where the square of the transverse gluon momentum $\vec{m}^2 << s = \vec{r}^2$ and the square of the longitudinal gluon momentum $z^2 \sim O(\vec{m}^2/\vec{r}^2)$, the non-factorized cross section $d\sigma/dQ^2$ depends on

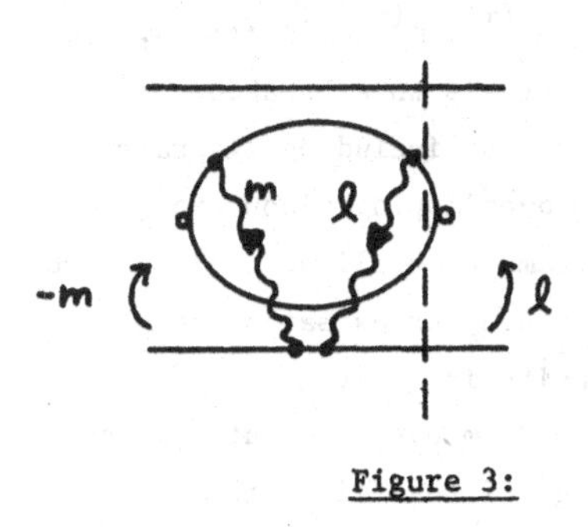

<u>Figure 3:</u>

$$\bar{\omega} = \bar{\omega}_B \frac{\alpha_s^2}{\pi^2} \hat{C} \int \frac{dz}{z} d\vec{m} \int \frac{dy}{y} d\vec{l}\, \psi_2^* \psi_1^* \hat{N}\, \Sigma\, \psi_1 \psi_2$$

where for this cut the color factor is $\hat{C} = C_F^2 - \frac{1}{2} C_F C_{YM}$, the numerator factor $\hat{N} = [\vec{r} \cdot \vec{m}\, \vec{r} \cdot \vec{l}/yz + \ldots]$ (the omitted terms are not important in the IR limit of the on-shell Glauber region), and the energy denominator factor is Σ. In the $\lambda_c \phi^3$ model for the bound state wave functions for each Kinoshita diagram the gluon momenta can be routed such that these bound state wave functions (the ψ's) are the same for each single cut, as is the $\hat{N}$ factor. Since $\hat{N}$ omits terms which are important when the gluons are in their UV regions, in analyzing this dynamical region we cannot shift gluon integration variables.

To study the soft-IR region as a limit of one-parameter, we replace the $(z, \vec{m})$ gluon variables with (z, μ, ϕ) gluon variables by defining $\vec{m} = (z - M)\vec{\mu}$ for a gluon of mass $M \neq 0$ so in the massless case $m_\mu = z(1, \vec{\mu}^2, \vec{\mu})$ and $z \to 0$ specifies the soft-IR region. The above single-cut contribution is logarithmically divergent both in z and in y in the near on-shell Glauber region. For both l and m transverse gluons soft, as a consequence of the associated Kinoshita diagram identity, a summation over the five possible single-cuts shows that the C_F^2 contribution from this Kinoshita diagram vanishes (this has also been proven for an arbitrary number of real and virtual, interpion gluon ladder exchanges). However, the center single-cut has no $C_F C_{YM}$ contribution and so, unlike in QED, there is a breakdown of the Kinoshita Diagram Identity in QCD for initial state spectator interactions. The simplest, though unphysical, example is the $C_F C_{YM}$ contribution for two instantaneous gluons of Fig. 4. A similar breakdown is found for other longitudinal $l^{(+)}$, $m^{(+)}$ directions. For g^4 virtuals this breakdown is not avoided by including diagrams with overlapping gluon exchanges. Diagrams with all soft gluons and a 3 gluon vertex satisfy the identity in order g^4.

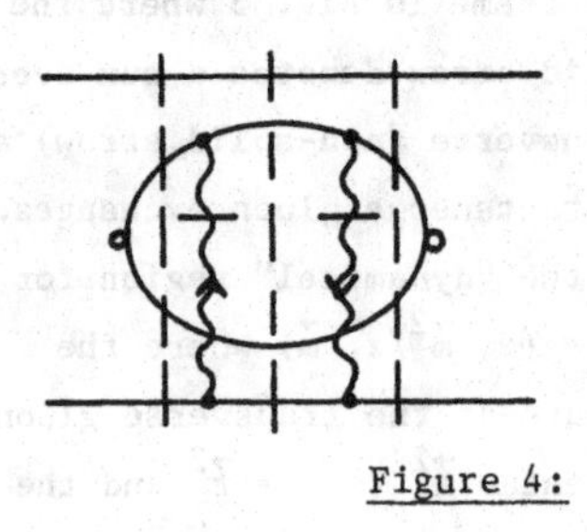

Figure 4:

Next we consider what happens when we "sum over spectator interactions" -- that is, we consider all the g^4 virtual diagrams with non-overlapping ladder exchanges in the near on-shell Glauber region.

There are 4 choices for $m^{(+)}$, $l^{(+)}$ directions and 16 possible non-overlapping diagrams for each. These can be grouped into subsets each of which contain 4 diagrams, e.g. a subset with an [AS] type m gluon and a right F cut is Fig. 5 where the overall sign of the $C_F C_{YM}$ color

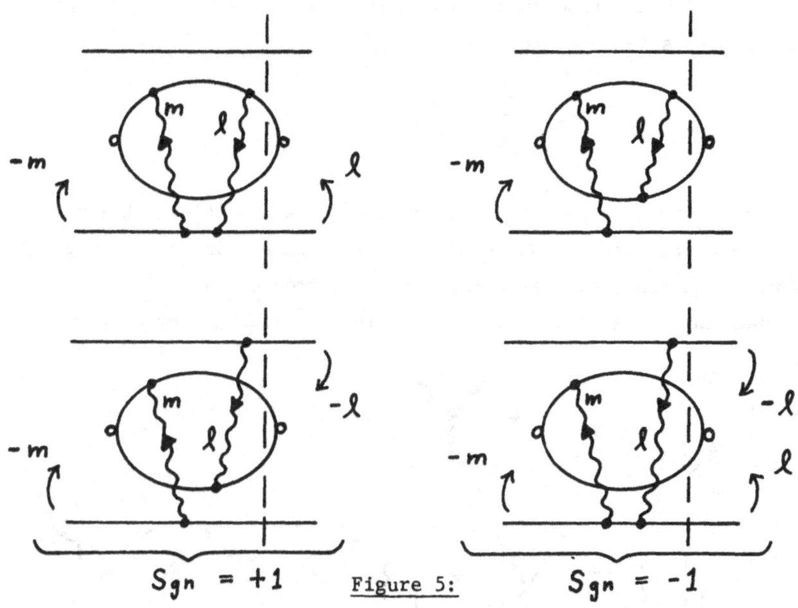

Figure 5:

factor is opposite for the diagrams in the left column to that for those in the right. These right F-type cuts give the same Σ and $\hat{N}$ for each diagram so for all soft gluons, the subset vanishes. Also when the outside gluon (the l gluon in this case) is soft, the subset vanishes! In this manner the "candidate" can be shown to pass the outside soft-IR test! (The K-type cuts always pass thru one of the outside gluons and so consequently, they always cancel.)

However, the contribution for the above subset of 4 diagrams depends on

$$\overline{R} = \int \frac{dy}{y} \int \frac{dz}{z} \frac{1}{\vec{k}_1^{\,2}\;\overline{k_2+z\mu}^{\,2}} \left[\frac{1}{\vec{k}_1^{\,2}} - \frac{1}{\overline{k_2-y\,\lambda}^{\,2}}\right]\left[\frac{1}{\overline{k_2+y\,\lambda}^{\,2}} - \frac{1}{\vec{k}_2^{\,2}}\right]$$

so when the inside m-gluon is soft, but the outside l-gluon is not soft, there is an uncancelled logarithmic soft-IR divergence. This divergence is not removed by summing over the 4 choices for $l^{(+)}$, $m^{(+)}$ directions. This is unphysical since a divergence from an interior exchange corresponds to phenomena from a small region of space whereas a long wavelength gauge quantum should sense only the large scale features of the physical phenomena. Although this particular difficulty is not present in the Abelian limit, inside soft-IR divergences are encountered in a straightforward evaluation of diagrams in QED. If per QED,[14] we introduce an identically conserved, $m^\mu g_\mu = 0$, internal gluon emission operator by

$$\gamma_\mu = g_\mu + R_\mu, \quad R_\mu = \frac{2P_\mu - m_\mu}{-m^2 + 2P \cdot m}$$

where P is the reference momenta for the quark or anti-quark line: then (i) there are no inside soft-IR divergences, and no leading-twist non-factorization effects, resulting from g_μ, and (ii) the non-Abelian matrix identity,[13] Fig. 6 in covariant gauge, suggests

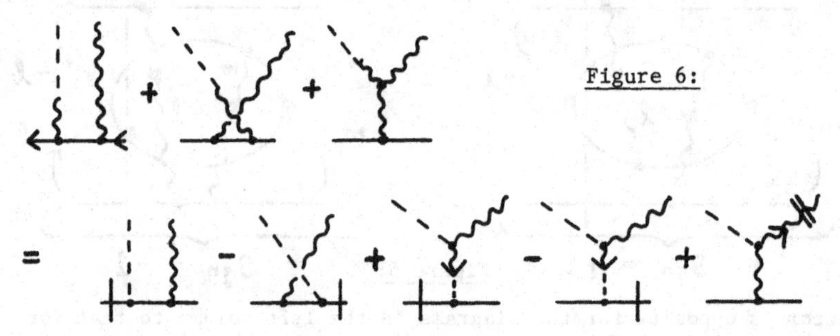

Figure 6:

3 gluon vertex diagrams must be included with the R_μ contributions, and then combined with the remainders arising from the removal of soft-IR factors from external legs.

Again, as for the spurious non-factorizing examples discussed at the beginning of this talk, gauge invariance is forcing consideration of a larger set of diagrams. In the case of this candidate involving virtual spectator quark-active quark interactions, Lindsay, Ross, and Sachrajda[12] have shown (and Bodwin, Brodsky, and Lepage[15]) in a covariant gauge that this g^4 apparent non-factorizing effect <u>also disappears</u> when diagrams involving the 3 gluon vertex are included and when parton distribution functions are carefully defined relative to spectator interactions in deep inelastic charged lepton-hadron scattering.

Taken together, the analysis of these examples suggests that if such a non-factorizing effect were to be experimentally discovered to exist at the S-matrix or cross-section level, then in QCD the discovery would imply the possibility of physically measuring the relative color phase of a local field operator at two distinct points in space-time.

REFERENCES

1. R. Doria, J. Frenkel and J. C. Taylor, Nucl. Phys. B169, 93 (1980); C. DiLieto, S. Gendron, I. G. Halliday and C. T. Sachrajda, Nucl. Phys. B183, 223 (1981); C. E. Carneiro, M. Day, J. Frenkel, J. C. Taylor and M. T. Thomas, Nucl. Phys. B183, 445 (1981).
2. C. A. Nelson, Nucl. Phys. B186, 187; B181, 141 (1981); D. R. Butler and C. A. Nelson, Phys. Rev. D18, 1196 (1978).
3. A. Andrasi, M. Day, R. Doria, J. Frenkel and J. C. Taylor, Nucl. Phys. B182, 104 (1981); I. Ito, Prog. Theor. Phys. 65, 1466 (1981); 67, 1216 (1982); N. Yoshida, Prog. Theor. Phys. 66, 269, 1803 (1981).
4. R. Fukuda, Phys. Lett. 73B, 33 (1978), and see references in Y. Kazama (these Proceedings).
5. T. Kinoshita, J. Math. Phys. 3, 650 (1962); T. D. Lee and M. Nauenberg, Phys. Rev. 133B, 1549 (1964); E. C. Poggio and H. R. Quinn, Phys. Rev. D14, 578 (1976); G. Sterman, Phys. Rev. D14, 2123 (1976); and G. Sterman and S. Weinberg, Phys. Rev. Lett. 39, 1436 (1977). Compare J. Cornwall and G. Tiktopoulos, Phys. Rev. D15, 2937 (1977); T. Kinoshita and A. Ukawa, Phys. Rev. D15, 1596 (1977); T. Appelquist and J. Carazzone, Nucl. Phys. B120, 77 (1977) and references in G. Sterman, A.I.P. Conf. Proc. No. 74 (ed. D. W. Duke and J. F. Owens, 1981).
6. J. Frenkel, J. G. M. Gatheral and J. C. Taylor, Nucl. Phys. B194, 172 (1982).
7. C. A. Nelson, SUNY BING 10/18/82 (to appear in Nucl. Phys.).
8. V. Ganapathi and G. Sterman, Phys. Rev. D23, 248 (1981).
9. T. Muta and C. A. Nelson, Phys. Rev. D25, 2222 (1982).
10. G. T. Bodwin, S. J. Brodsky and G. P. Lepage, Phys. Rev. Lett. 47, 1799 (1981); SLAC-PUB-2927 (1982).
11. A. H. Mueller, Phys. Lett. 108B, 355 (1982) and CU-TP-232; J. Collins, D. Soper and G. Sterman, Phys. Lett. 109B, 388 (1982) and ITP-SB-82-46; G. Date, ITP-SB-82-47.
12. W. Lindsay, D. Ross and C. Sachrajda, Phys. Lett. 117B, 105 (1982) and SHEP-81/82-6.
13. C. A. Nelson, SUNY BING 11/25/82 (to appear in Nucl. Phys.).
14. D. R. Yennie, S. C. Frautschi and H. Suura, Ann. Phys. (N.Y.) 13, 379 (1961); G. Grammer and D. R. Yennie, Phys. Rev. D8, 4332 (1973). Compare, S. Libby and G. Sterman, Phys. Rev. D18, 3252 (1978); J. C. Collins and G. Sterman, Nucl. Phys. B185, 172 (1981); J. C. Collins and D. E. Soper, OITS 155; and A. Sen, Phys. Rev. D24, 3281 (1981).
15. S. J. Brodsky (these Proceedings).

APPENDIX

SOME SAMPLE CONTEST ENTRIES

(1) In addition to the statement that the correct group is less than SU_{512}, it seems to indicate that Oklahoma oil is escaping to Kansas.

(2) Sooner dead than red. I'm okay, you're okay, Oklahoma's okay. Oil makes the world go round.

(3) Oil makes it possible to bring color to the Big Eight.

(4) The SU(3) theory of 3 colors of quarks and 8 gluons, if solved nonperturbatively may have practical applications.

(5) The logo shows that Oklahoma used to extend much further to the south but that sometime in the past, someone, probably a Texan, ripped it off. To make amends, they sent a scaled-down version of the Eiffel tower.

(6) The logo illustrates the well-known panhandle diagram which plays a central role in the deconfinement of quark oil plasma in the high temperature phase of the standard triple orthogonal SU_8 model.

(7) 0 = oil, also symbolizes zero color states, which are the only ones which can be freely extracted from the ground state.
8 = middle day of the conference, hence day around which conference schedule is reflection symmetric; also smallest nontrivial reflection symmetric representation of SU_3; there are 8 subsidiary conditions in SU_3.
3 = number of days in conference; also month of the year; also fundamental representation of SU(3).
83 = year of conference.
SU(3) = Obvious to anyone attending!

(8) High energy physicists should not be allowed to dream up inscrutable logos.

LIST OF PARTICIPANTS

NAME	INSTITUTION
Adler, Stephen	Institute for Advanced Study
Andrew, Keith	Arkansas
Ball, James	Utah
Bambah, Bindu	Chicago
Bender, Carl	Washington U.
Breit, John	Institute for Advanced Study
Brodsky, Stanley	SLAC
Carroll, Jeff	OSU
Chang, F. C.	St. John's
Cornwall, J.M.	UCLA
Creutz, Michael	Brookhaven
Dolan, Louise	Rockefeller
Fiegel, Robert	Oklahoma
Ghneim, Said	OSU
Greenberg, O. W.	Maryland
Grose, Ted	OSU
Haymaker, Richard	LSU
Hokim, Q.	Laval University
Hudson, Bruce	Oklahoma
Johnson, Kenneth	MIT
Johnson, Porter	IIT
Joseph, David	Nebraska
Kalbfleisch, George	Oklahoma
Kantowski, Ronald	Oklahoma
Karnal, Arif	Oklahoma
Kazama, Yoichi	Kyoto
Kohler, Gary	Oklahoma
Koppel, Thomas	Chalk River
Kruecken, Thomas	Nebraska

NAME	INSTITUTION
Lasher, Gordon	IBM
Laursen, Morten	OSU
Lee, H. C.	Chalk River
Lieber, Michael	Arkansas
McKay, Douglas	Kansas
Milton, Kim	OSU
Morgan, Thomas	Nebraska
Munczek, Herman	Kansas
Nelson, Charles	SUNY at Binghamton
Nemirovsky, Adolfo	Kansas
Pinsky, Stephen	Ohio State
Polonyi, Janos	Illinois
Reid, James	Simon Fraser
Rno, Jung	Cincinnati
Rutledge, D. L.	OSU
Samuel, Mark	OSU
Sanganetra, Pahol	Kansas
Saritepe, Seljuk	OSU
Schierholz, Gerrit	Hamburg
Sen, Achin	OSU
Shakin, Carl	Brooklyn College
Sivers, Dennis	Argonne
Skubic, Pat	Oklahoma
Soni, Amarjit	UCLA
Summers, Geoffrey	OSU
Swamy, N. V. V. J.	OSU
Sylvester, Garrett	OSU
Thomaz, Maria Teresa	Wisconsin
Tupper, Gary	OSU
Wilcox, Walter	OSU

NAME	INSTITUTION
Williams, P. K.	DOE
Willis, Sue	Oklahoma
Wilson, Timothy	OSU
Woloshyn, Richard	TRIUMF

(OSU = Oklahoma State University)

Progress in Mathematics

Edited by J. Coates and S. Helgason

Progress in Physics

Edited by A. Jaffe and D. Ruelle

- A collection of research-oriented monographs, reports, notes arising from lectures or seminars
- Quickly published concurrent with research
- Easily accessible through international distribution facilities
- Reasonably priced
- Reporting research developments combining original results with an expository treatment of the particular subject area
- A contribution to the international scientific community: for colleagues and for graduate students who are seeking current information and directions in their graduate and post-graduate work.

Manuscripts

Manuscripts should be no less than 100 and preferably no more than 500 pages in length.

They are reproduced by a photographic process and therefore must be typed with extreme care. Symbols not on the typewriter should be inserted by hand in indelible black ink. Corrections to the typescript should be made by pasting in the new text or painting out errors with white correction fluid.

The typescript is reduced slightly (75%) in size during reproduction; best results will not be obtained unless the text on any one page is kept within the overall limit of 6x9½ in (16x24 cm). On request, the publisher will supply special paper with the typing area outlined.

Manuscripts should be sent to the editors or directly to:
Birkhäuser Boston, Inc., P.O. Box 2007, Cambridge, Massachusetts 02139

PROGRESS IN MATHEMATICS

Already published

PM 1 Quadratic Forms in Infinite-Dimensional Vector Spaces
Herbert Gross
ISBN 3-7643-1111-8, 432 pages, paperback

PM 2 Singularités des systèmes différentiels de Gauss-Manin
Frédéric Pham
ISBN 3-7643-3002-3, 346 pages, paperback

PM 3 Vector Bundles on Complex Projective Spaces
C. Okonek, M. Schneider, H. Spindler
ISBN 3-7643-3000-7, 396 pages, paperback

PM 4 Complex Approximation, Proceedings, Quebec, Canada, July 3-8, 1978
Edited by Bernard Aupetit
ISBN 3-7643-3004-X, 128 pages, paperback

PM 5 The Radon Transform
Sigurdur Helgason
ISBN 3-7643-3006-6, 207 pages, hardcover

PM 6 The Weil Representation, Maslov Index and Theta Series
Gérard Lion, Michèle Vergne
ISBN 3-7643-3007-4, 348 pages, paperback

PM 7 Vector Bundles and Differential Equations
Proceedings, Nice, France, June 12-17, 1979
Edited by André Hirschowitz
ISBN 3-7643-3022-8, 256 pages, paperback

PM 8 Dynamical Systems, C.I.M.E. Lectures, Bressanone, Italy, June 1978
John Guckenheimer, Jürgen Moser, Sheldon E. Newhouse
ISBN 3-7643-3024-4, 305 pages, hardcover

PM 9 Linear Algebraic Groups
T. A. Springer
ISBN 3-7643-3029-5, 314 pages, hardcover

PM 10 Ergodic Theory and Dynamical Systems I
A. Katok
ISBN 3-7643-3036-8, 346 pages, hardcover

PM 11 18th Scandinavian Congress of Mathematicians, Aarhus, Denmark, 1980
Edited by Erik Balslev
ISBN 3-7643-3040-6, 526 pages, hardcover

PM 12 Séminaire de Théorie des Nombres, Paris 1979-80
Edited by Marie-José Bertin
ISBN 3-7643-3035-X, 404 pages, hardcover

PM 13 Topics in Harmonic Analysis on Homogeneous Spaces
Sigurdur Helgason
ISBN 3-7643-3051-1, 152 pages, hardcover

PM 14 Manifolds and Lie Groups, Papers in Honor of Yozô Matsushima
Edited by J. Hano, A. Marimoto, S. Murakami, K. Okamoto, and H. Ozeki
ISBN 3-7643-3053-8, 476 pages, hardcover

PM 15 Representations of Real Reductive Lie Groups
David A. Vogan, Jr.
ISBN 3-7643-3037-6, 776 pages, hardcover

PM 16 Rational Homotopy Theory and Differential Forms
Phillip A. Griffiths, John W. Morgan
ISBN 3-7643-3041-4, 258 pages, hardcover

PM 17 Triangular Products of Group Representations and their Applications
S. M. Vovsi
ISBN 3-7643-3062-7, 142 pages, hardcover

PM 18 Géométrie Analytique Rigide et Applications
Jean Fresnel, Marius van der Put
ISBN 3-7643-3069-4, 232 pages, hardcover

PM 19 Periods of Hilbert Modular Surfaces
Takayuki Oda
ISBN 3-7643-3084-8, 144 pages, hardcover

PM 20 Arithmetic on Modular Curves
Glenn Stevens
ISBN 3-7643-3088-0, 236 pagers, hardcover

PM 21 Ergodic Theory and Dynamical Systems II
A. Katok, editor
ISBN 3-7643-3096-1, 226 pages, hardcover

PM 22 Séminaire de Théorie des Nombres, Paris 1980-81
Marie-José Bertin, editor
ISBN 3-7643-3066-X, 374 pages, hardcover

PM 23 Adeles and Algebraic Groups
A. Weil
ISBN 3-7643-3092-9, 138 pages, hardcover

PM 24 Enumerative Geometry and Classical Algebraic Geometry
Patrick Le Barz, Yves Hervier, editors
ISBN 3-7643-3106-2, 260 pages, hardcover

PM 25 Exterior Differential Systems and the Calculus of Variations
Phillip A. Griffiths
ISBN 3-7643-3103-8, 349 pages, hardcover

PM 26 Number Theory Related to Fermat's Last Theorem
Neal Koblitz, editor
ISBN 3-7643-3104-6, 376 pages, hardcover

PM 27 Differential Geometric Control Theory
Roger W. Brockett, Richard S. Millman, Hector J. Sussmann, editors
ISBN 3-7643-3091-0, 349 pages, hardcover

PM 28 Tata Lectures on Theta I
David Mumford
ISBN 3-7643-3109-7, 254 pages, hardcover

PM 29 Birational Geometry of Degenerations
Robert Friedman and David R. Morrison, editors
ISBN 3-7643-3111-9, 410 pages, hardcover

PM 30 CR Submanifolds of Kaehlerian and Sasakian Manifolds
Kentaro Yano, Masahiro Kon
ISBN 3-7643-3119-4, 223 pages, hardcover

PM 31 Approximations Diophantiennes et Nombres Transcendants
D. Bertrand and M. Waldschmidt, editors
ISBN 3-7643-3120-8, 349 pages, hardcover

PM 32 Differential Geometry
Robert Brooks, Alfred Gray, Bruce L. Reinhart, editors
ISBN 3-7643-3134-8, 267 pages, hardcover

PM 33 Uniqueness and Non-Uniqueness in the Cauchy Problem
Claude Zuily
ISBN 3-7643-3121-6, 185 pages, hardcover

PM 34 Systems of Microdifferential Equations
Masaki Kashiwara
ISBN 0-8176-3138-0
ISBN 3-7643-3138-0, 182 pages, hardcover

PM 35 Arithmetic and Geometry Papers Dedicated to I. R. Shafarevich on the Occasion of His Sixtieth Birthday Volume I Arithmetic
Michael Artin, John Tate, editors
ISBN 3-7643-3132-1, 373 pages, hardcover

PM 36 Arithmetic and Geometry Papers Dedicated to I. R. Shafarevich on the Occasion of His Sixtieth Birthday Volume II Geometry
Michael Artin, John Tate, editors
ISBN 3-7643-3133-X, 495 pages, hardcover

PM 37 Mathématique et Physique
Louis Boutet de Monvel, Adrien Douady, Jean-Louis Verdier, editors
ISBN 0-8176-3154-2
ISBN 3-7643-3154-2, 454 pages, hardcover

PM 38 Séminaire de Théorie des Nombres, Paris 1981-82
Marie-José Bertin, editor
ISBN 0-8176-3155-0
ISBN 3-7643-3155-0, 359 pages, hardcover

PM 39 Classical Algebraic and Analytic Manifolds
Kenji Ueno, editor
ISBN 0-8176-3137-2
ISBN 3-7643-3137-2, 644 pages, hardcover

PM 40 Representation Theory of Reductive Groups
P. C. Trombi, editor
ISBN 0-8176-3135-6
ISBN 3-7643-3135-6, 308 pages, hardcover

PM 41 Combinatorics and Commutative Algebra
Richard P. Stanley
ISBN 0-8176-3112-7
ISBN 3-7643-3112-7, 102 pages, hardcover

PM 42 Théorèmes de Bertini et Applications
Jean-Pierre Jouanolou
ISBN 0-8176-3164-X
ISBN 3-7643-3164-X, 140 pages, hardcover

PROGRESS IN PHYSICS

Already published

PPh 1 Iterated Maps on the Interval as Dynamical Systems
Pierre Collet and Jean-Pierre Eckmann
ISBN 3-7643-3026-0, 256 pages, hardcover

PPh 2 Vortices and Monopoles, Structure of Static Gauge Theories
Arthur Jaffe and Clifford Taubes
ISBN 3-7643-3025-2, 294 pages, hardcover

PPh 3 Mathematics and Physics
Yu. I. Manin
ISBN 3-7643-3027-9, 112 pages, hardcover

PPh 4 Lectures on Lepton Nucleon Scattering and Quantum Chromodynamics
W.B. Atwood, J.D. Bjorken, S.J. Brodsky, and R. Stroynowski
ISBN 3-7643-3079-1, 574 pages, hardcover

PPh 5 Gauge Theories: Fundamental Interactions and Rigorous Results
P. Dita, V. Georgescu, R. Purice, editors
ISBN 3-7643-3095-3, 406 pages, hardcover

PPh 6 Third Workshop on Grand Unification, 1982
P.H. Frampton, S.L. Glashow, H. van Dam, editors
ISBN 3-7643-3105-4, 388 pages, hardcover

PPh 7 Scaling and Self-Similarity in Physics
(Renormalization in Statistical Mechanics and Dynamics)
J. Fröhlich, editor
ISBN 3-7643-3168-2
ISBN 0-8176-3168-2, 440 pages, hardcover